T0181699

SpringerBriefs in Business

More information about this series at http://www.springer.com/series/8860

Gerardo Marletto · Simone Franceschini
Chiara Ortolani · Cécile Sillig

Mapping Sustainability Transitions

Networks of Innovators, Techno-economic Competences and Political Discourses

Gerardo Marletto
DiSEA
University of Sassari
Sassari
Italy

Simone Franceschini
DiSEA
University of Sassari
Sassari
Italy

Chiara Ortolani
DICEA
University of Rome "La Sapienza"
Rome
Italy

Cécile Sillig
DIEM
University of Genova
Genoa
Italy

ISSN 2191-5482 ISSN 2191-5490 (electronic)
SpringerBriefs in Business
ISBN 978-3-319-42272-5 ISBN 978-3-319-42274-9 (eBook)
DOI 10.1007/978-3-319-42274-9

Library of Congress Control Number: 2016944845

Printed on acid-free paper

This Springer imprint is published by Springer Nature
The registered company is Springer International Publishing AG Switzerland

To Giacomo and Nicola, you don't need a map, just follow your nature!

Preface

With this book, we have set ourselves an ambitious target to provide the readers with a simple graphical tool—the socio-technical map—that may help them understand how the environmental sustainability of human activities may be reached. In other terms, we have tried to make simple a very complex process. To reach this goal, we founded our work on the recently emerged research field of sustainability transitions (SusTrans). The starting point of the SusTran approach is that most of human needs are fulfilled by systems which have proved to be environmentally unsustainable, and it is precisely for this reason that these systems should be radically changed. In other words, system innovation is at the heart of SusTrans.

Systems that fulfill human needs feature two relevant characteristics which should be seriously taken into consideration when aiming at the ambitious goals of curbing greenhouse gas emissions and reducing all other environmental impacts of human activities. First characteristic: These systems are embedded in the very structure of our society. Behind each good and service we produce and consume every day, one can find several interconnected elements: individuals and organizations, values and ideologies, technologies and infrastructures, markets and industries, rules and norms, and so forth. Thus, the functioning of these systems simultaneously involves the cultural, institutional, technological, and economic dimensions of human life: any attempt at reductionism—for example, to the technological or economic dimension of such systems—is not correct. Second characteristic: These systems are continuously changing. Their dynamics is generated by a process of structured action: systems are changed by human agents, but their action is in turn conditioned by the very structure of systems. Moreover—and maybe more important here—the outcome of such a process is uncertain: no change is known ex-ante, but it emerges as the (sometimes unintended) result of human action. This is why any attempt to describe system innovation as the result of a choice between given alternatives is not correct. If we keep in mind these two characteristics, we realize that an effective approach to environmental sustainability must be: (1) systemic, that is, able to consider all dimensions of innovation, not only technologies or markets, but also behavior, rules, discourses, policies, etc.; (2) dynamic, that is, able to understand how new systems emerge through the alignment of innovations as they are generated.

In order to meet these two requirements, the SusTran approach mostly refers to the socio-technical (ST) analysis of innovations. Here, any societal function is fulfilled by one or more ST-systems, and its evolution is strongly affected by the interaction between a stable and dominant ST system and other existing or emerging ST-systems. Each of these ST-systems is supported by a network of agents interested in its reproduction. Following ST theories and concepts, a SusTran can be represented as a ST transition to sustainability: a process of radical change where the current dominant position of an unsustainable ST system is destabilized and took over by a new—and environmentally sustainable—ST system. In such a representation, innovators play a key role: both those innovating to protect a dominant position and those innovating to take it over. Such a competition between alternative ST-systems—and their supporting networks of innovators—is strongly affected by policies. In particular, a ST system is holding a dominant position also because it was able to gain support from favorable policies that now protect the status quo. This is why a SusTran may not take place without new and specific policies that: (a) help emerging ST-systems to consolidate and (b) destabilize the dominant ST system. This in turn implies that the competition between alternative networks of innovators taking place along a SusTran is not only about technological leadership and market power; it also refers to the ability of influencing public debates, political agendas, and actual policies. In other terms, political innovators and political innovations are relevant features of any SusTran.

Summing up: SusTrans are complex processes. They feature systemic changes, structured action, innovation-driven dynamics and—last but not least—competition on multiple dimensions. As stated at the beginning of this Preface, the ambitious goal of this book is to provide the reader with a simple graphic tool—the socio-technical map (ST-map)—which can be used to represent a SusTran without renouncing to the understanding of its complexity. In particular, with this book any scholar or practitioner interested to the issue of sustainability will be able to understand how the current situation of any societal function may give place to different ST transition pathways. The dynamics of the relevant networks of innovators—and their technological, market, and political strategies—will be the key variables to assess the likelihood of the resulting alternative scenarios. The ST-map can also be used to understand which policies are deemed necessary to support those ST transition pathways—if any—that can be considered as SusTrans.

The book is divided as follows: In Chap. 1, the essential basic concepts that are needed to build a ST-map will be presented; a specific attention will be given to the definition of ST-systems and SusTrans and to the role played by network of innovators. In Chap. 2, we will explain in detail how to build a ST-map and how to use it to generate alternative ST transition pathways; in particular, we will see how to position in the ST-map all relevant networks of innovators and how to forecast their future dynamics. In Chaps. 3–6, the results of some case studies will be provided: the societal functions of feeding, mobility, and lighting were analyzed in order to ascertain if the ST-map can be adapted to different situations. With the case of food, we verified that the ST-map is able to represent a transition featuring different (if not conflicting) visions of sustainability; with the cases of mobility, we found that both global and local dynamics can be considered starting from the same ST-map; in the case of lighting, the ST-map proved useful to generate new knowledge on the overall dynamics of dominant ST-systems and policies. On the whole, case studies also confirmed that the ST-map is a more appropriate and powerful analytic tool for prospective rather than retrospective studies.

Sassari, Italy
Rome, Italy
Genoa, Italy
May 2016

Gerardo Marletto
Simone Franceschini
Chiara Ortolani
Cécile Sillig

Acknowledgements

I thank Anthony Doyle, Senior Editor at Springer, for asking me to submit a book proposal, starting from the paper "Car and the city: socio-technical transition pathways to 2030" published in 2014 by *Technological Forecasting & Social Change*. In this paper, the socio-technical map was proposed for the first time as a graphical tool to represent sustainability transitions.

I am glad to thank the colleagues that authored the case studies that integrate this book and can help the reader to understand how the socio-technical map can be used in practice: Simone Franceschini for the case of lighting and Chiara Ortolani for the case of mobility in Freiburg (D). Cécile Sillig co-authored with me the case of food.

I would also thank Isfort (the Italian higher institute for research and training in transportation) for the co-funding of Simone and Cécile past research activities.

Contents

Abbreviations

ACS	Automatic control system
CFL	Compact fluorescent lamp
ESCO	Energy-saving company
EV	Electric vehicle
GE	General Electric
HP	Hewlett Packard
LED	Light-emitting diode
NPP	Nuclear power plant
OLED	Organic light-emitting diode
SG	Smart grid
ST	Socio-technical
ST-map	Socio-technical map
SusTran	Sustainability transition
WTO	World Trade Organization

About the Authors

Gerardo Marletto is an associate professor of applied economics at the University of Sassari (I). His research interests are institutional and evolutionary economics; sustainability economics and policy; and participation in public decisions. He has published several papers in academic journals and edited the book "Creating a sustainable economy" (2012). E-mail: marletto@uniss.it.

Simone Franceschini holds a Ph.D. at the Technical University of Denmark on eco-innovation dynamics in the lighting sector, with a specific interest in the systemic impact of eco-efficiency innovations. Currently, he is a fellow researcher at University of Sassari, with a project on developing public participation for decision-making processes. He holds a M.Sc. in Economics at the University of Rome (I) "Tor Vergata" and a M.Sc. in Innovation and Knowledge at the Aalborg University (DK).

Chiara Ortolani holds a Ph.D. in urban planning and carries out teaching and research activities at the Living the City Lab of the Department of Civil, Building and Environmental Engineering (DICEA)—University of Rome (I) "La Sapienza." Her research interests focus on cycling and pedestrian mobility, with a specific reference to the issues of energy consumption, air pollution, urban morphology, and public space planning.

Cécile Sillig is a research fellow at University of Genoa (I), Department of Economics. She holds a master's degree in Geography and a Ph.D. in Transport Economics. Her research interests are port regions development, logistics city networks, sustainable food transportation, and innovations for sustainable agribusiness.

Part I
The Socio-technical Map

Chapter 1
Basic Concepts

Abstract In this chapter the essential basic concepts that are needed to build a socio-technical map are presented. Any social function—such as feeding, housing, mobility, supply of energy, healthcare, etc.—is fulfilled by one or more socio-technical (ST) systems. Each ST system consists of a network of innovators and a structure of material and immaterial constituents. Usually one ST system holds a dominant position: only ST "niches" are partially or totally protected from its selection pressure. The dynamics of ST systems may be grouped into two large families: the adaptation of a dominant ST system and the establishment of a new dominant position. Niches play a relevant role in both kinds of dynamics. The political dimension of the dynamics of ST systems becomes relevant when a dominant position is taken over: niche innovators must scale up a cumulative process between empowerment, legitimation and networking in order to gain a stable role into the public debate and possibly to influence the direction of change of agendas and actual policies. A sustainability transition (SusTran) is needed when a social functions is currently dominated by (or locked into) a ST system that is environmentally unsustainable. The mere adaptation of an existing dominant ST system is not sufficient to generate a SusTran. This is why the take-over of the dominant position of an unsustainable ST system is a necessary condition for a SusTran to take place.

Keywords Socio-technical system · Socio-technical niche · Innovators · Transition pathways · Sustainability transition

1.1 The Socio-technical Analysis of System Innovations

The seminal book of Geels (2005) is the basic reference to understand both the theoretical foundations and the potential applications of the socio-technical (ST) analysis of system innovation. The explicit consideration of both the social and the technological sides of system innovation processes is based on a very simple statement: "technologies do not fulfil societal functions on their own"

G. Marletto et al., *Mapping Sustainability Transitions*,
SpringerBriefs in Business, DOI 10.1007/978-3-319-42274-9_1

(ibidem: 1). To understand any system innovation process one must look at the strict interaction between human actions and social structures that make a technology functional. All the material and immaterial elements mobilised by such a strict interaction constitute a "ST system".

In this vision "system innovation is a transition from one socio-technical system to another" (ibidem: 2). But only ex-post a system innovation can be represented as a mere substitution between an old and a new ST system; if one look at a transition before (or while) it deploys what he/she will see it is rather a competition between alternative (and more or less stable and powerful) ST systems. As we will see, such a competition takes place along all the interacting dimensions of a transition (markets, technologies, political institutions) and it is implemented by conflicting networks of innovators (each supporting a specific ST system).

1.2 Socio-technical Systems: Structure and Agency

Any social function—such as feeding, housing, mobility, supply of energy, healthcare, etc.—is fulfilled by one or more ST systems. Each ST system is a (a more or less) stable configuration consisting of a network of supporting social agents and a structure of material and immaterial constituents (infrastructures, knowledge, rules, financial resources, etc.). Social agents who perform a more active role to support a ST system are also called "core actors"; other less relevant supporting agents are also called "fringe actors" (Smith et al. 2005).

Any ST system cover all the relevant dimensions of a social function; in particular, three main sub-systems can be found in a ST system: markets, technologies and political institutions. All sub-systems are key to make a system innovation process viable: markets play a relevant role in coordinating all economic decisions that are taken during an out-of-equilibrium process (Amendola and Gaffard 2006); technologies are made of artefacts and routinised knowledge, and their change is driven by a collective heuristic (a "paradigm") (Dosi 1982); political institutions provide the formal and tacit background to foster social experimentation and to avoid social sclerosis (North 2005). Sub-systems interact, thus generating bi-lateral and trilateral dynamic relations and assuring the overall consistency of the ST system while it changes.

Most social agents are organizations—such as firms, public authorities, NGO, research bodies, etc.—that replicate and change the structure of the ST system. In particular, they generate—intentionally or unintentionally, directly or indirectly—the variation and selection of the economic, technological and political constituents of the ST system they belong to. The ability of a social agent to influence the dynamics of a ST system—and the whole social function—is a positive function of his/her material and immaterial endowments (physical and financial resources, knowledge and skills, social capital and legitimacy, etc.). In more general terms, the complex dynamics of any ST system can be conceptualised as structured agency: its structure is replicated and changed through an iterative process of action and learning of individual and collective social agents, which in turn is enabled and constrained by the structural—and interacting—constituents of the ST system. Thus the functioning of a ST system is genuinely dynamic and path dependent: future changes are neither completely uncertain (as the potential developments of the existing ST systems are limited by its structure) nor completely certain (as the interaction of structural variables and agency may generate unpredictable outcomes). Actually, the dynamics of ST systems is more than uncertain, it is non-ergodic, as even changes in the fundamental structure of a ST system are usual features of its dynamics. In such a context it is no longer possible to consider agency as driven by complete information: on the contrary, agents' rationality is bounded and adaptive, and their behaviour is not optimizing, but satisfying (Simon 1987).

ST systems may feature different levels of stability and power. Stability—or, more exactly, homeorhesis—of a ST system is mainly driven by a core set of rules which is shared by its supporting agents and influences the direction of change. This set is also called the "regime" of the ST system (Geels 2002). It must be stressed that, just as the whole ST system covers the technological, market and political dimensions of a social function, its regime is made not only of techno-economic routines and heuristics, but also of socio-political ideologies and discourses. Usually one ST system holds a dominant position, that is, it is very stable and strongly influences the dynamics of the whole social function, and of other stable (but subaltern or residual) ST systems. In particular, dominant positions usually generate path-dependence and lock-in phenomena at the level of the whole societal function which in turn hinder the emergence of new ST systems. Only ST "niches" are partially or totally protected from the selection pressure generated by the dominant ST system. Taking advantage of any kind of barrier (geographical, technological, commercial, institutional), ST niches are essential for the incubation and experimentation of innovations, and for the gradual structuring and empowerment of a new ST system. Before that possibly happens, ST niches feature—by definition—low levels of both stability and power (Schot and Geels 2007; Smith and Raven 2012).

Most of the above concepts may be expressed in terms of Darwinian mechanisms. In particular: (a) the dynamic interactions between the markets, technologies and political institutions of a ST system could be described as co-evolutionary processes, and (b) path-dependence and lock-in phenomena could be expressed in terms of repeated selection of the dominant ST system (Safarzynska 2012).

1.3 Networks of Innovators and Their Demography

As stated above, a network of supporting social agents is a relevant constituent of a ST system. These social agents must be considered as innovators whenever they are interested in changing the ST system they belong to. Even when a ST system is defending its dominant position through innovation its core actors must be considered as innovators. Social agents that are interested into the emergence of a new ST system should always be considered as innovators; these innovators—often starting their activities in ST niches—are also called "enactors" (Suurs et al. 2010).

If we remember that a ST system is made of three main sub-systems (markets, technologies and political institutions), then we should acknowledge that a network of innovators spanning on the economic, technological and political dimensions of a social function is an essential requirement for a ST system to change consistently. This means that innovators contribute not only to technological or commercial novelties, but also to changes taking place in the political dimension of a social function. As a consequence, also social agents proposing new issues into the public debate and into the political agenda must be considered as innovators; such innovators are also called "social entrepreneurs" (Brown and Vergragt 2012) or "cultural entrepreneurs" (Zweynert 2009). Following this dynamic representation, the regime of any ST system can be considered as the set of routinized multi-dimension innovation strategies of its supporting network.

There is a strict relation between agents' power and their ability to generate effective political innovations. In particular, core actors of a dominant ST system feature high levels of power and legitimacy and they are able to use their endowments to influence the dynamics of politics and policy. On the contrary, enactors must scale up a cumulative process between empowerment, legitimation and networking in order to gain a stable role into the public debate and possibly to influence the direction of change of agendas and actual policies (Avelino and Rotmans 2009). This is why some scholars describes the competition between networks of dominant core-actors and networks of enactors as a "battle over institutions" or—in order to stress that social agents' interests and political narratives are entrenched—as a "battle over discourses" (Hekkert et al. 2007; Kern 2011).

A multilevel selection framework can be used to represent the resulting co-dynamics between networks of innovators and political institutions. Through an upward agency mechanism, networks of innovators with different interests and visions use their power to influence political institutions, thus affecting the overall dynamics of a whole social function. Through a downward structuration mechanism—and as a result of changes in political institutions—power is differentially bestowed to networks of innovators, and their further evolution is affected. As a consequence, a cumulative causation process may take place: the more a network of innovators is able to influence political institutions and increase its resources, the more it attracts other social agents and networks (bringing along their resources), the more that network is able to influence the political institutions and increase its resources, and so on. Such a process may reach a tipping point—and generate the lock-in of the whole social function—when a network of innovators is able to gain support from the dominant policy.

It is also apparent that the whole "demography" of networks of innovators is relevant to understand the dynamics of a social function: the creation of a new network from scratch; an individual agent joining a network or migrating from a network to another; the merging, splitting and re-assortment of networks; etc. Inter

Table 1.1 A socio-technical representation of the car system

Social function		Urban mobility
Structural constituents	Markets	Oligopolistic barriers to entry: huge investments on car plants and advertising
	Technologies	Internal combustion
		Networks of roads and highways
		Other networks (gas stations, repair stations, etc.)
	Political institutions	Rules to allocate urban and suburban space to the car
		Public provision of road networks
		Incentives to buy 'green' cars
Core actors		Car industries
		Oil industries
		Road industries
		National and federal Authorities
Competing ST systems and niches		Public transport
		Bicycle
		Sharing schemes
Routinized strategies (regime)	Markets	Sell cars to individuals
	Technologies	Increase the efficiency of internal combustion (most car manufacturers)
		Hybridize the internal combustion and electric propulsion (some car manufacturers)
	Political institutions	Lobby to gain support (or low pressure) from national/federal policies
	Networking	Absorption as core actors of the producers of batteries

alia, all the above implies that the dynamics of a social function is generated by both cooperative and competition mechanisms, taking place within and between networks of innovators, respectively. The relevance of network changes in system innovation processes also makes clear that also networking strategies should be considered as a constituent of systems' regimes.

See Table 1.1 for a simplified ST representation of the car system.

1.4 Transition Pathways: Adaptation Versus Take-Over

The dynamics of ST systems may be grouped into two large families: the adaptation of a dominant ST system and the establishment of a new dominant position. Adaptation can be conceptualized as a homeorhetic process: changes in institutions, markets and technologies are driven by the existing regime and take place along an established transition pathway; the alignment of such changes is granted by the structure of the dominant ST system—which gradually changes—and it is supported by its core actors.

Things completely change when a ST system try to gain the dominant position: a process of extrication is needed to free resources, knowledge, actors, etc., that are locked into the dominant system; intentional and unintentional forces generating lock-in and path-dependence phenomena must be overcome; new institutions, technologies and markets must be built; a new process of multidimensional alignment must be triggered and the resulting ST transition made viable. But neither a regime nor a structure are available to coordinate all these efforts: both the regime and the structure are created through the innovation process. This is why the establishment of a new dominant position is an exceptional event which is virtually impossible without the increasingly coordinated action of enactors.

Niches play a relevant role in both kinds of dynamics: in the case of adaptation, niches may cluster with the dominant ST system; in the case of the establishment of a new dominant position, niches contribute to threaten the dominant ST system and possibly take it over.

Geels and Schot (2007) have provided a typology of ST transition pathways in which the role of innovators is explicitly considered. Haxeltine et al. (2008) explain such a typology in terms of "transformative mechanisms" that allow innovators to have access to new material and immaterial endowments through the creation or reconfiguration of their networks. As shown in Table 1.2, four transition pathways can be considered. "Transformation" occurs when core actors gradually adjust a dominant ST system in order to respond to external pressures for change; these pressures can be both top-down (e.g., caused by new policies) and bottom-up (e.g., generated by grassroots movements). In this case innovation is mostly incremental and adaptive and—especially when new innovative competences are needed—it may be possibly supported by the absorption into the dominant ST system of some enactors, or even the clustering of their whole niches. "Reconfiguration" takes place when core actors are able to respond to external or internal pressures by partially changing the regime and the structure of the dominant ST system, in particular by creating a permanent link to some enactors and to the innovations they developed in one or more niches. "Substitution" is the result of a 'battle': core actors—possibly coming from other social functions or other geographical areas—profit from the pressures on the dominant ST system and—after taking it over—radically change it. "De-alignment and re-alignment" involve enactors coming from one or more ST niches and subaltern ST systems that, whilst the dominant ST system is destabilized by major external pressures, gradually create a new network of innovators and eventually establish a new dominant ST system. In both the transition pathways that lead to a take-over, it may happen that some core actors of the previously dominating ST system are absorbed into the new emerging network of innovators.

See Table 1.2 for an overview of ST transition pathways and some examples taken from alternative scenarios of 2030 urban mobility (Chap. 4 of this book).

Table 1.2 Socio-technical transition pathways: an overview with examples from the case of urban mobility[a]

Overall dynamics	Adaptation of the dominant ST system		Creation of a new dominant position (takeover)	
Transition pathway	Transformation	Reconfiguration	Substitution	De-alignment and re-alignment
Innovators' main strategy	Core actors react to external pressures	Integration of new actors into the supporting network of innovators	Core actors of other ST systems take over and change the dominant ST system	A network of enactors establishes a new ST system while the dominant ST system is destabilized by external pressures
Main transformative mechanisms	Internal adjustment and maintenance	Absorption of new actors	Competition between the dominant ST system and a new ST system	Clustering and empowering of niches and subaltern systems
Examples from 2030 scenarios of urban mobility	Core actors of the car system implement all innovations to make internal combustion more efficient	Producers of batteries entering the network of innovators of the car system in order to make the mass production of electric cars possible	Creation of a new system of urban mobility supported by innovators coming from the electric sector that integrates renewables, smart grids and electric vehicles	Creation of a new system of urban mobility that integrates all systems and niches which are alternative to the car (public transport, carsharing, bicycle)

[a]Adapted from Geels and Schot (2007), Haxeltine et al. (2008) and Chap. 4 of this book

1.5 The Relevance of Political Innovations in Takeovers

In the case of the adaptation of a dominant ST system the economic and technological dimension of the ST transition pathway are more relevant: innovators who are able to implement commercial and technological innovations are the main drivers of change. The political dimension is less relevant, yet active. Core actors of the dominant ST system invest their endowments: (a) to keep gaining support (or a weak pressure) from the dominant approach to policy, and (b) to counteract the voice of enactors and core actors of other ST system into the public debate. If necessary, the dominant network may try to absorb some opposing or competing social agents in order to: benefit from their pressure for innovation; weaken their potential disruptive effects; and avoid the risk that they coalesce with others (Walker 2000).

The political dimension of the ST transition pathway becomes more relevant than the economic and technological ones when a dominant position is taken over. The emergence of a new—and potentially dominant—network of innovators results from a cumulative process that may be triggered by one or more of the following factors: the migration of a stable ST system from another societal function; the coalescing of many niches and subaltern ST systems; the increasing empowerment of a ST system that—after reaching a dominant position in a single location—migrates to other geographical areas. With the exception of the former case, techno-economic innovators may not have a leading role since the beginning: only when business opportunities become apparent they start playing a more active role. In all other cases grassroots political innovators play a relevant role, especially in the starting steps of the transition. Afterwards, socio-political and techno-economic enactors realize that their actions and goals are consistent and may be coordinated; this why a network of innovators is gradually created that is able to scale up the cumulative causation process between the enlistment of an increasing number of members and the growing influence on political institutions. At the beginning of this process political legitimation is the main target, then explicit advocacy and direct lobbying become more and more important, also with the purpose to destabilize the existing dominant position. Before achieving durable credibility and a stable influence on agendas, formal norms and policies, the emerging network must be able to affect shared cultures, political discourses and ideas, and informal rules. In other words, political innovation is a key driver of a ST take-over. When successful, the ST transition pathway reaches a tipping point and ends up with the whole societal function locked in a new ST system, whose dominant position is supported by new dominant policies.

1.6 Sustainability Transitions as Socio-technical Takeovers

A sustainability transition (SusTran) is nothing but the ST transition of a social function towards environmental sustainability. A SusTrans is needed when a social functions is currently dominated by (or locked into) a ST system that is environmentally unsustainable (Unruh 2000, 2002). It is widely acknowledged by the relevant literature that the mere adaptation of an existing dominant ST system is not sufficient to generate a SusTran (van den Bergh et al. 2011; Markard et al. 2012). This why the take-over of the dominant position of an unsustainable ST system—and the process of radical and multidimensional innovation it brings along—is a necessary condition for a SusTran to take place. In particular, the competition between alternative networks of innovators is a key feature of any SusTran and—especially at the beginning of the ST transition pathway—it is also played on the political dimension of the whole social functions. As stated above, on one side, core actors of the currently dominating ST system keep implementing their political strategies to prevent and counteract any possible pressure coming from existing and potential competing ST systems; on the other side, enactors promote new political discourses and political routines in order to generate those new agendas and actual policies that are needed to support the emergence and empowering of a new and more sustainable ST system that may eventually gain a dominant position.

1.7 The Urban/Local Level of Transition Pathways: A Note

ST systems and SusTrans are usually analyzed at a national/international level. Sometimes the urban/local level is taken into account, but just as a recipient for the implementation of an innovation process generated at a higher scale. Only in recent years scholars have acknowledged that at the urban/local level of transition pathways: (a) coalitions of enactors can be built more easily; (b) local endowments may be mobilized for innovative practices; (c) political deliberation is more fluid (Bulkeley et al. 2011). But—as clearly stated by Geels (2011)—the city can feature a more relevant role than the mere hosting of niches: locally dominant ST systems may co-exist with another ST systems that dominates the societal functions at the national/international level; ST niches may be located at the local/urban level, but then the dynamics of the dominant system takes place at the national/international level. Also as a result of these dynamics, multi-scalar processes must be taken into consideration: both innovators and innovations may not only migrate from a local area to another, but they can also "travel" across spatial levels, from the local to the global, and back (Sengers and Raven 2015).

References

Amendola M, Gaffard JL (2006) The market way to riches: behind the Myth. Edward Elgar, Cheltenham and Northampton

Avelino F, Rotmans J (2009) Power in transition. An interdisciplinary framework to study power in relation to structural change. Eur J Soc Theory 12:543–569

Brown H, Vergragt PJ (2012) Grassroots innovations and socio-technical system change–Energy retrofitting of the residential housing stock. In: Marletto G (ed) Creating a sustainable economy. Routledge, Abingdon

Bulkeley H, Castan Broto V, Hodson M, Marvin S (2011) Cities and low carbon transitions. Routledge, Abingdon

Dosi G (1982) Technological paradigms and technological trajectories—a suggested interpretation of the determinants and directions of technical change. Res Policy 11:147–163

Geels FW (2002) Technological transitions as evolutionary reconfiguration processes: a multi-level perspective and a case study. Res Policy 31:1257–1274

Geels FW (2005) Technological transitions and system innovations: a co-evolutionary and socio-technical analysis. Edward Elgar, Cheltenham
Geels FW (2011) The role of cities in technological transitions: analytical clarifications and historical examples. In: Bulkeley H, Castan Broto V, Hodson M, Marvin S (eds) Cities and low carbon transitions. Routledge, Abingdon
Geels WF, Schot J (2007) Typology of sociotechnical transition pathways. Res Policy 36:399–417
Haxeltine A, Whitmarsh L, Bergman N, Rotmans J, Schilperoord M, Köhler J (2008) A conceptual framework for transition modelling. Int J Innov and Sustain Develop 3:93–114
Hekkert M, Suurs RAA, Negro S, Kuhlmann S, Smits R (2007) Functions of innovation systems: a new approach for analysing technological change. Technol Forecast Soc Change 74:413–432
Kern F (2011) Ideas, institutions, and interests: explaining policy divergence in fostering 'system innovations' towards sustainability. Environ Plann C Govern Policy 29:1116–1134
Markard J, Raven R, Truffer B (2012) Sustainability transitions: an emerging field of research and its prospects. Res Policy 41:955–967
North D (2005) Understanding the process of economic change. Princeton University Press, Princeton and Oxford
Safarzynska K (2012) Agency and economic change. In: Marletto G (ed) Creating a sustainable economy. Routledge, Abingdon
Schot J, Geels FW (2007) Niches in evolutionary theories of technical change. A critical survey of the literature. J Evol Econ 17:605–622
Sengers F, Raven R (2015) Toward a spatial perspective on niche development: the case of bus rapid transit. Environ Innov Soc Transitions 17:166–172
Simon HA (1987) Models of man. Garland, New York and London
Smith A, Raven R (2012) What is protective space? reconsidering niches in transition to sustainability. Res Policy 41:1025–1036
Smith A, Stirling A, Berkhout B (2005) The governance of sustainable socio-technical transitions. Res Policy 34:1491–1510
Suurs R, Hekkert M, Kieboom S, Smits R (2010) Understanding the formative stage of technological innovation system development: the case of natural gas as an automotive fuel. Energy Policy 38:419–431
Unruh GC (2000) Understanding carbon lock-in. Energy Policy 28:817–830
Unruh GC (2002) Escaping carbon lock-in. Energy Policy 30:317–325
van den Bergh JCJM, Truffer B, Kallis G (2011) Environmental innovation and societal transitions: introduction and overview. Environ Innovation Soc Transitions 1:1–23
Walker W (2000) Entrapment in large technology systems: institutional commitment and power relations. Res Policy 29:833–846
Zweynert J (2009) Interests versus culture in the theory of institutional change? J Inst Econ 5: 339–360

Chapter 2
How to Build a Socio-technical Map

Abstract In this chapter we show how a socio-technical map (ST-map) can be built in four steps. Graphical examples are taken from the case studies presented in Chaps. 3–6 of this book. In step 1 the field of analysis is determined: this is usually a whole societal function, such as feeding, housing, mobility. In step 2 the relevant systems and niches and their networks of innovators are identified: the dominant system (if any), other systems, and niches are represented differently into the ST-map. Also local systems and niches—when relevant—can be represented in a specific way. In step 3 the two dimensions of the ST-map are drawned. The first dimension usually represents the relevant political discourses on sustainability, that is, how the current and future sustainability of the analyzed societal function is interpreted by innovators. The second dimension of the ST-map usually represents the techno-economic competences that are leveraged by innovators to promote change. These may refer either to business or productive models, or to technologies. In step 4 all ST systems and niches are positioned into the ST-map. Such a positioning represents to which political discourse and to which techno-economic competence mostly refers the network of innovators of any given system or niche. Also the dominant policy (if any) can be positioned into the ST-map. The resulting ST-map can be used to represent either the current situation of a societal function or its prospective or retrospective dynamics. In the latter case the history of a societal function can be represented through a sequence of ST-maps, each referring to a single moment of relative stability of the societal function. Starting from the current ST-map and from the analysis of the current potential for change also scenarios can be built: these are represented as the result of the 'demography' of ST systems and their supportive networks.

Keywords Socio-technical map · Socio-technical systems · Network of innovators · Political discourse · Scenario analysis

G. Marletto et al., *Mapping Sustainability Transitions*,
SpringerBriefs in Business, DOI 10.1007/978-3-319-42274-9_2

2.1 What Is the Socio-technical Map for?

As we have seen in the previous chapter, a sustainability transition (SusTran) is a complex process of system innovation involving both the techno-economic and socio-political dimensions of a societal function. Such a process is made possible by innovators that are able at the same time to foster cooperation (into a network supporting a ST system) and competition (between ST systems).

With the ST-map we pretend to provide the reader with a simple graphical tool that can be used to represent a SusTran without loosing the understanding of its complexity. We think that the ST-map can be used by scholars, practitioners and policy makers as a tool both for the analysis of current and future system innovations, and for the design and implementation of strategies for sustainability.

In the following of this chapter we show how a ST-map can be built. Some graphical examples are taken from the case studies presented in Chaps. 3–6. We refer readers to these chapters for a deeper understanding of how a ST-map can be built and used in real cases.

2.2 How to Build a Socio-technical Map in Four Steps

Step 1—Determine what is the field of analysis

The socio-technical (ST) analysis of SusTrans usually applies to societal functions, such as feeding, housing, mobility, etc. It is just at such an overall level that societal changes generated by system innovations can be more easily analyzed and understood. In particular, all relevant political changes that make SusTrans viable take place at the level of societal functions (if not at the level of the whole society). When a ST-map is applied to societal sub-functions its analytic potential is reduced; e.g. in the case of lighting proposed in Chap. 6, it is apparent the risk to focus on techno-economic innovations, while political changes are mostly considered exogenous.

Societal functions—and all systems and niches that provide it—usually reproduce on multiple scales, that is, simultaneously at the global, national and local level. Moreover, the dynamics at different scales may show some misalignment. The ST-map is able to represent such a space complexity by combining the global picture with national/local specificities (see below, Step 2). The study of mobility in the city of Freiburg (D) (Chap. 5 of this book) shows that the ST-map can also be used to analyze how a societal function changes in a specific local area. These very local ST-maps are useful to understand why local configurations (dominant positions; systems and niches; policies; etc.) differ from global ones.

Step 2—Identify the relevant systems and niches and their networks of innovators

Once determined which is the societal function to be analyzed, it is usually easy to identify all systems that contribute to its provision, and—for each of them—who

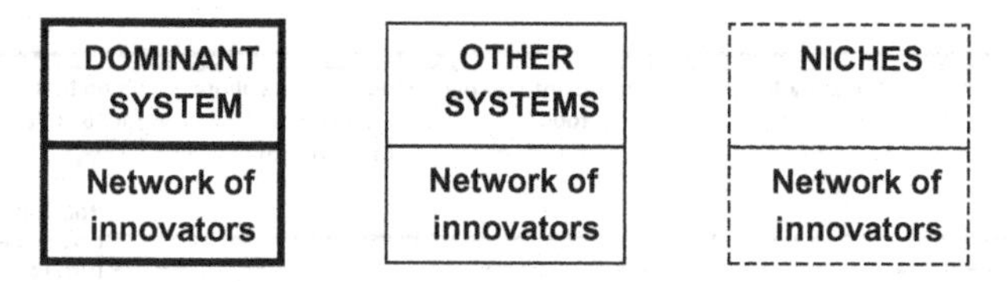

Fig. 2.1 Graphical representation of socio-technical systems and niches and their supporting network of innovators

are the main members of their supporting network of innovators. It must be remembered that innovators can be: Authorities, companies, other organizations (such as political or trade associations), grassroots movements, media, etc.

The dominant system (if any), other systems, and niches are represented differently into the ST-map (Fig. 2.1).

Also local systems and niches—when relevant—can be represented in a specific way. As shown in the examples taken from Chap. 4 (Figs. 2.2 and 2.3), in this case the graphical representation of the network of innovators is lost, and this information must be given in the accompanying text.

Step 3—Draw the two dimensions of the ST-map

The ST-map is framed by two dimensions.

Dimension 1 represents the relevant political discourses on sustainability, that is, how the current and future sustainability of the analyzed societal function is interpreted by innovators. The standard articulation of sustainability in its environmental, social and economic constituents should be remembered when looking for political discourses. Figure 2.4 provides examples taken from three of the following case studies to show how this dimension can be drawn: here political discourses are represented with a claim in order to make them more understandable to the reader.

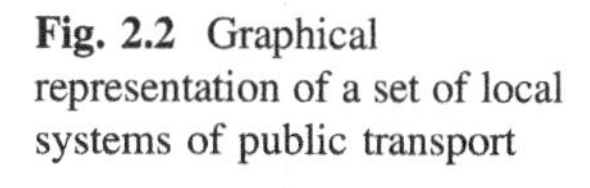

Fig. 2.2 Graphical representation of a set of local systems of public transport

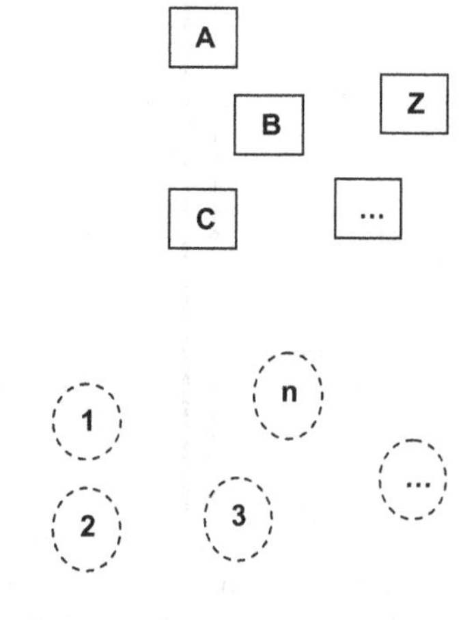

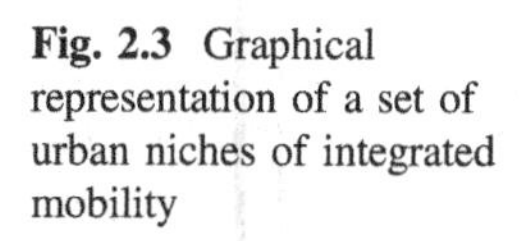

Fig. 2.3 Graphical representation of a set of urban niches of integrated mobility

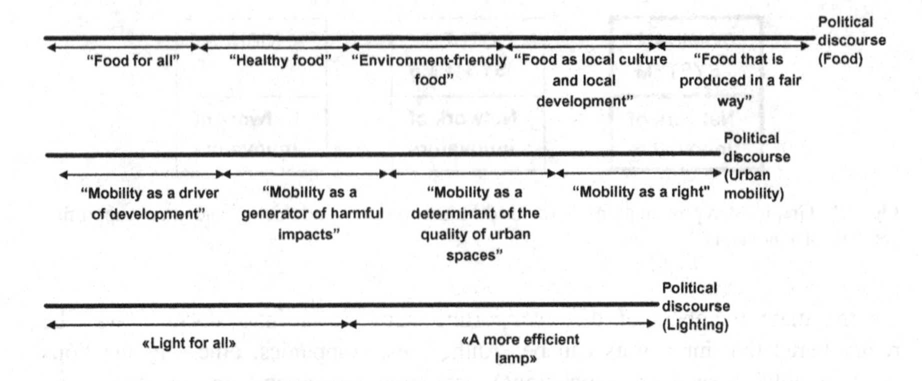

Fig. 2.4 Graphical representation of the dimension 'Political discourse'. Examples taken from the case studies on food, urban mobility and lighting

Dimension 2 of the ST-map represents the techno-economic competences that are leveraged by innovators to promote change. As shown in the examples taken from the case studies presented in the following chapters, such competences may refer either to business or productive models (Fig. 2.5). In other cases, technologies may be considered.

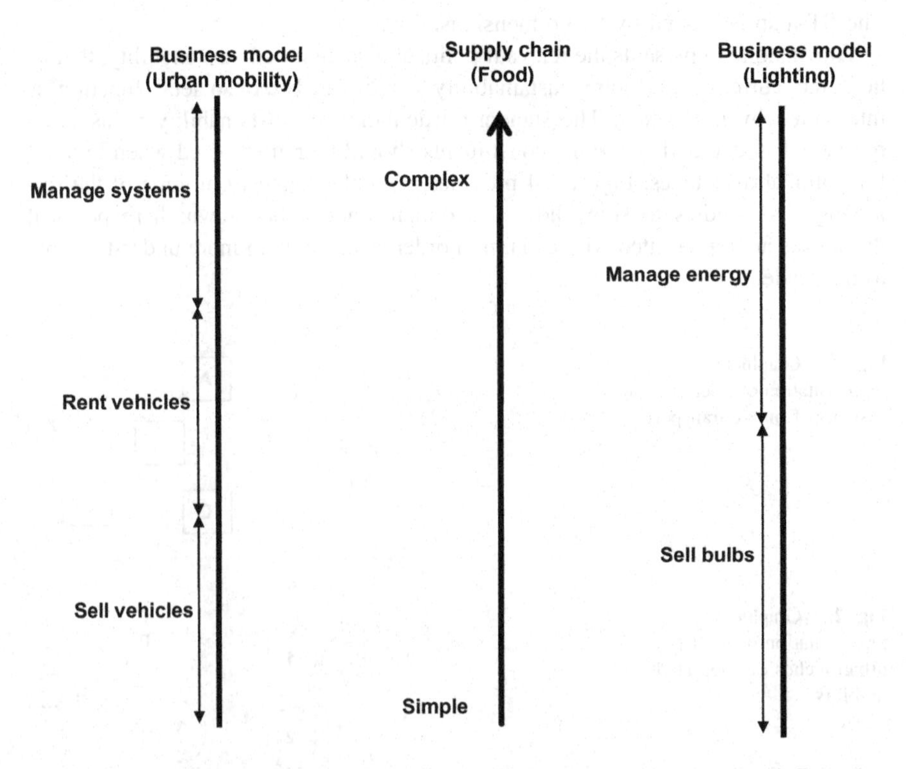

Fig. 2.5 Graphical representation of the dimension 'Techno-economic competences'. Examples taken from the case studies on food, urban mobility and lighting

Step 4—Position systems, niches (and the dominant policy) into the ST-map

Systems and niches previously identified can now be positioned with reference to the two dimensions of the ST-map. Such a positioning represents to which political discourse and to which techno-economic competence mostly refers the network of innovators of any given system or niche.

Figure 2.6 represents the ST-map of the current situation of urban mobility (taken from Chap. 4). This example clearly shows that some systems (such as the individual car) are unequivocally centered on one political discourse and on one techno-economic competence; instead, other systems may refer to two political discourses (such as the individual bicycle) or to two techno-economic competences (such as some local sharing schemes). Things become more difficult to represent when a systems range between more than two competences or political discourses; a solution to this problem is provided in the ST-map of the societal function of feeding where most systems are able to manage supply chains featuring different levels of complexity (Fig. 2.7; taken from Chap. 3).

Also the dominant policy can be represented into the ST-map. Its positioning must be interpreted as that of systems and niches. It must be stressed that when the dominant policy is inside a system (usually the dominant one) that means that a networks of innovators has been able to influence (and possibly to "capture") the political dimension of the whole societal function. See again Figs. 2.6 and 2.7 for graphical examples.

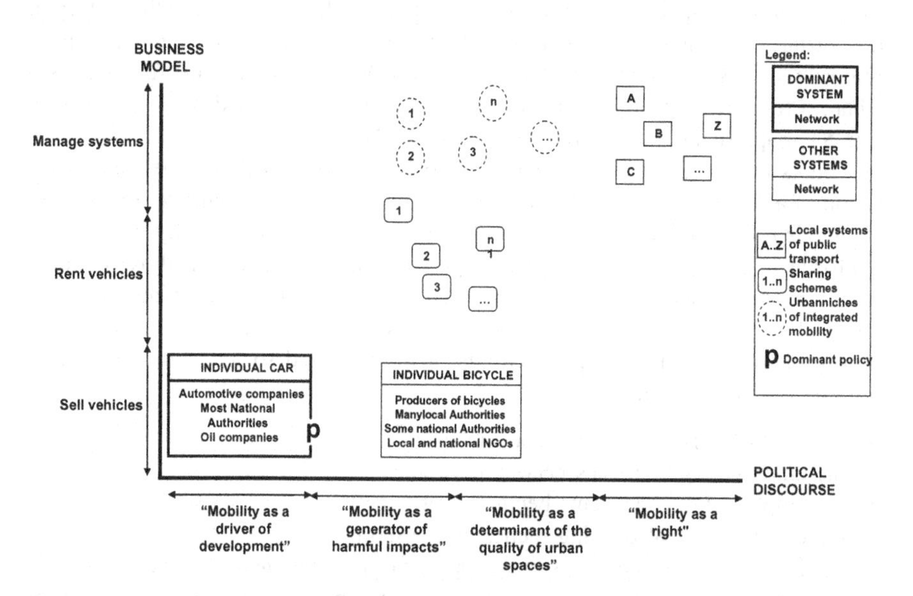

Fig. 2.6 The socio-technical map of urban mobility: current situation

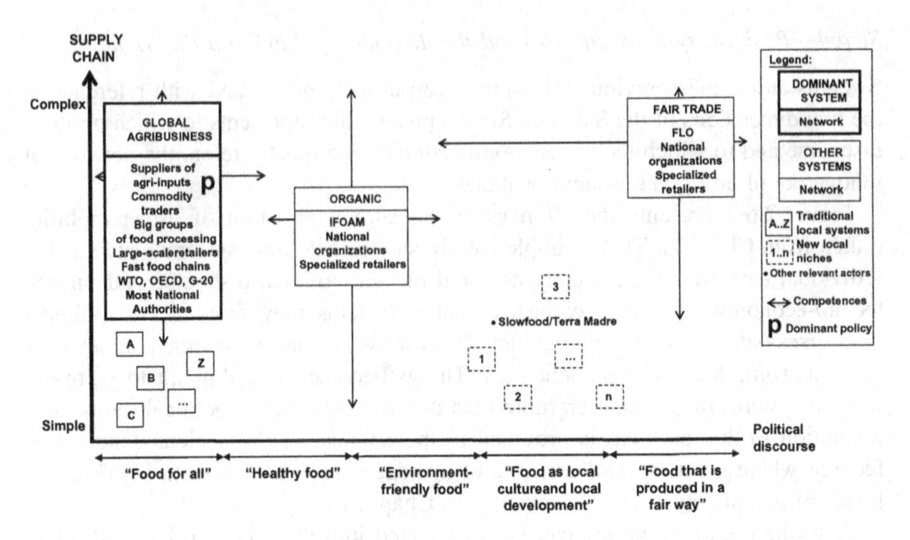

Fig. 2.7 The socio-technical map of feeding: current situation

2.3 From Maps to Transitions

The ST-map can be used to represent either the current situation of a societal function or its prospective or retrospective dynamics. In the latter case the history of a societal function can be represented through a sequence of ST-maps, each referring to a single moment of (relative) stability of the societal function. See Chap. 6 for a retrospective analysis of the lighting sector based on a sequence of five ST-maps spanning over the period 1880–2010.

Things are more complex when ST-maps are used to envisage the future evolution of a societal function. In these cases, the analysis of the current potential for change is used to build (usually more than one) scenario. Consistently with the ST approach to system innovation, such a potential is to be found in those ongoing techno-economic and political changes that can trigger and make viable the 'demography' of systems and their supportive networks. In particular, the result of the transition from current to future settings of the analyzed societal function can be represented in the ST-map as:

- the shift of existing systems and niches,
- the empowering of niches becoming systems,
- the empowering of systems becoming dominant,
- the disappearing of existing niches and systems,
- the destabilization (and possible disappearing) of dominant systems,
- the emergence of new niches and systems (possibly reaching a dominant position),
- the disappearing or emergence of a dominant policy,
- the entering of systems from outside the societal function,

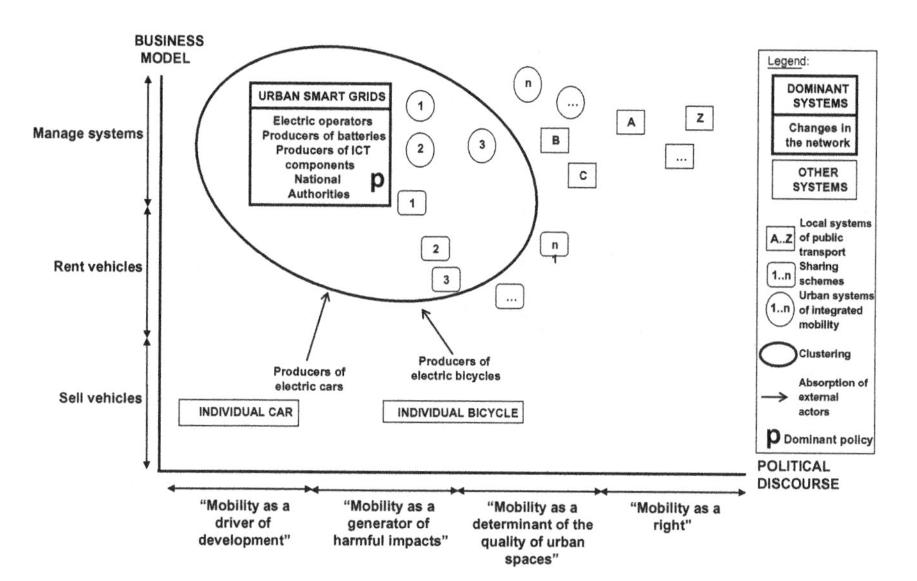

Fig. 2.8 The socio-technical map of urban mobility: 2030 'Electricity' scenario

- the absorption of new members in a network of innovators (possibly coming from other systems or societal functions),
- the clustering of (either existing or new) systems and niches.

Figure 2.8 (taken from Chap. 4) represent one of the three alternative 2030 scenarios of urban mobility. Here the reader can see how the dynamics of absorption and clustering can be represented in a ST-map.

Part II
Case Studies

Chapter 3
Towards Sustainable Food: A Contested Transition

Abstract The aim of this chapter is to show that the socio-technical map (ST-map) can be useful even to analyze the current and future dynamics of feeding: a societal function that feature complex techno-economic dynamics and contrasting political visions. First of all, the ST systems that are relevant for the today provision of food are considered: the global agribusiness system; the traditional and new local systems; the organic system; the fairtrade system. For each of them a supporting network of innovators is identified; all relevant policies—at the global, national and local scale—are also considered. Then, all food ST systems are positioned into a ST-map based on two dimensions. The first dimension is built on five alternative discourses on sustainable food: 'Food for all', 'Healthy food', 'Environment-friendly food', 'Food as local culture and local development', and 'Food that is produced in a fair way'. The other dimension ranks supply chains from simple to complex: complexity of supply chains depends on both the distance covered and the number of steps. The chapter ends with a future analysis of sustainable food. Starting from ongoing changes and from alternative dynamics of ST-systems and networks of innovators, three scenarios are proposed. The 'Biotech' scenario results from a two-fold transition involving the global agribusiness system: suppliers of biotechnological inputs respond to the pressure for sustainable agriculture and food security; large-scale retailers integrate the commercial activity of the organic, fairtrade and local systems. The 'Elite Versus Mass' scenario features a new network led by elitist large-scale retailers that clusters the commercial activities of the organic and fair trade systems in order to serve the increasing demand for healthy and environment-friendly food. In the 'Organic' scenario a new dominant system integrates all the productive and commercial actors who are interested in converting to organic practices and to organic food products; also some large-scale retailers coming from the global agribusiness system join the organic network. In all scenarios the cumulative process between the empowerment of networks and the emergence of supporting policies is stressed.

Keywords Food · Agribusiness · Network of innovators · Sustainability transition · Socio-technical map · Scenario analysis

G. Marletto et al., *Mapping Sustainability Transitions*,
SpringerBriefs in Business, DOI 10.1007/978-3-319-42274-9_3

3.1 Introduction

The aim of this case-study is to show that the socio-technical map (ST-map) can be useful even in the case of societal functions that feature complex techno-economic dynamics and contrasting political visions.

Feeding is one of such cases. Indeed, the provision of food is backed by several vertically linked industries (production of agri-inputs, agriculture, industrial transformation, retailing) each featuring specific internal dynamics. The resulting productive and commercial systems are both global and local. Moreover the sustainability of food is interpreted in multiple (and to some extent conflicting) ways. As a result, several political approaches to the sustainability of food are at stake, while different (and sometimes inconsistent) policies are implemented at multiple geographical scales (urban/local, national/federal, multilateral/global).

In this case the relevant literature will be used to build the ST-map of feeding and the resulting sustainability pathways and scenarios. In particular:

- ST food systems and niches will be positioned with reference to the techno-economic and sociopolitical cognitive elements that are crucial for any future innovation;
- for each ST food system and niche a supporting network of innovators will be identified;
- co-evolutionary dynamics between such networks and ongoing changes will be used to build alternative transition pathways.

3.2 Sustainable Food Today: A Global Issue

3.2.1 Who's Feeding the World? Food Systems, Actors and Supporting Policies

Most of the food consumed by the world population is provided by two different kinds of productive and commercial systems (Green et al. 2003):

- the global agribusiness system. This mostly covers developed and emerging economies and include highly industrialized agriculture (with an increasing use of GMOs-genetically modified organisms[1]), industrial food processing, large scale retailing and modern consumption patterns. These latter include the high- and increasing-percentage of industrially processed food in the daily diet;

[1] Almost 90 % of the hectarage of GMOs crops is concentrated in five countries: USA (40.4 % of the world total), Brazil (23.3 %), Argentina (13.4 %), India (6.,4 %), Canada (6.4 %) (Clive 2014).

- a number of traditional local systems. These are mostly diffused in developing and poor countries and include labour-intensive agriculture, traditional retailing and the consumption of non-transformed products. Also subsistence agriculture and home-made transformed food can be part of these systems. These systems have been increasingly weakened by global liberalization policies that have brought along the elimination of import–export regulations and tariffs protecting domestic productions. As a result, also developing and poor countries are more and more served by the global agribusiness system.

The global agribusiness system is dominated by corporations that mostly operates in the following world oligopolistic markets that feature medium to high levels of concentration (Patel 2003; Murphy 2006): (a) provision of inputs for agriculture (e.g., Monsanto, Dow Chemical-DuPont and ChemChina-Syngenta), (b) trading of agricultural commodities (e.g., Cargill, ADM, Louis Dreyfus, Bunge and Glencore), (c) industrial transformation of agricultural commodities into food products (e.g., Nestlè, Pepsico, Unilever and Mondelez), and (c) selling of final products in large-scale retailing chains (such as Walmart, Carrefour, Tesco, Metro and Kroger). It must be stressed that some big transformers also control the production and/or the brokering of agricultural commodities (e.g., Cargill). Some of these global corporations are among the largest world companies. E.g.: Walmart ranks 1st in the Fortune 500 list and—with its 2 millions employees—is by far the world largest employer; other agrifood corporations rank in the top-50 of Fortune 500: Kroger (20th), ADM (34th), Pepsico (44th), Coca Cola (63rd); also some fast food and retailing chains are among the world largest employers, such as Tesco (with 520,000 employees), McDonald's (420,000), Carrefour (365,000), Yum Brands (300,000), Starbucks (190,000).[2] Agrifood corporations feature not only market power, but also political power, as they are able to gain support from both national/federal and global/transnational trade policies (Clapp and Fuchs 2009). In particular: large (but decreasing) subsidies are granted to the agriculture of developed economies, such as the USA and Canada, the EU, Japan and South Korea, Australia and New Zealand (Anderson et al. 2013; Brooks 2014); the world production and trade of agricultural commodities and food products have been strongly affected by liberalization and deregulation policies fostered by supranational institutions, in particular by the initiatives of the World Trade Organization (WTO 1995, Annex 1A; WTO 2001, artt. 13 and 14).

The overall picture of the global provision of food is completed by two other systems and a set of emerging niches that are alternative to the global agribusiness system.

Organic agriculture is a system based on a set of techniques that avoid the use of synthetic compounds (fertilizers, pesticides, animal drugs and food additives) and largely rely on ecological processes and cycles adapted to local conditions (FiBL and IFOAM 2016). It must be stressed that two thirds of the global organic

[2]Data sources: Fortune 500, 2015 edition; UNCTAD, statistics on transnational corporations, 2012.

agriculture land is concentrated in Oceania (39.7 % of the world total) and Europe (26.6 %); and only in these two continents the share of certified organic on total agriculture land is not negligible: 4.1 % in Oceania and 2.4 % in Europe (while is near to zero in Africa and Asia) (FiBL and IFOAM 2016). This system is backed and reproduced by a multitude of actors; most of them operate at a local or national scale and their certified products are sold in specialized shops. But some organic actors are able to manage international supply chains that integrate both specialized and large-scale retailing chains. NGOs and single operators of more than 100 countries are part of the IFOAM–Organics International world association, founded in 1972 in order to coordinate the actions of all members (advocacy and lobbying; research and studies; conferences and forums; trade fairs and exhibitions; accreditation and certification; etc.). The organic system benefits of national supporting policies that are implemented in particular in those countries with a relevant share of organic agriculture (Austria, Italy, Switzerland, etc.); in some cases pro-organic local policies are implemented too (e.g., for the use of organic food in schools, hospitals and nursing homes).

The so-called 'fair trade' of food products is a system organized by several organizations with the aim of ensuring equitable prices and incomes to local producers. Such producers are usually located in poor, developing or emerging countries, and their products are usually sold in developed countries; that implies that fair trade food systems incorporates long (and sometimes, very long) supply chains. Since the mid-80's, fair trade organizations have given birth to second-level international networks, such as: EFTA–European Fair Trade Association; FLO-I–Fair Labelling Organizations International; IFAT–International Federation of Alternative Trade, (now WFTO–World Fair Trade Organization); NEWS–Network of European Worldshops. A world informal association of the above networks (called FINE) was created in 1998. All fair trade associations and networks mostly operated in the fields of standardization, certification, monitoring; advocacy and campaigning are implemented too, but not as core activities.

More and more new local systems are emerging in developed countries as the result of several initiatives: local/regional brand'ing, farmers' markets, community supported agriculture (CSA) and urban agriculture, food co-ops and buying groups, movements for local food and short(er) supply chains (Goodman et al. 2012). Most of these are grassroots actions, but an increasing number of them is supported by (mostly local) policies. These local systems are not coordinated into national or supra-national networks; an important exception is the international Slow food movement that involves local communities (called "convivia") promoting good, local, traditional food. Slow food was created in Italy in 1986 and since then has promoted many initiatives, among which worth remembering the "Terra Madre" (Mother Earth) global network (with participants from both the global North and South) and its world meetings, and the University of Gastronomic Sciences based in Bra (I).

Worth signaling that the societal function of feeding has also increasingly interacted with that of the provision of energy. Indeed, some agricultural commodities (e.g., soya, maize, palm, rapeseed, sugarcane) can be used for the production of food (for animals and humans) or for the production of biofuels.[3] The dynamics of this interaction is strongly conditioned by the prices of energy and agriculture commodities, both experiencing wide up and downs during the last ten years, also because of the increasing "financialization" of commodity global markets (Randall Wray 2008). Moreover, the US, EU and Brazil policies for renewable fuels have affected the world market of these agro-energy commodities (Al-Riffai et al. 2010; Bellemare and Carnes 2015).

3.2.2 *Alternative Discourses on Sustainable Food*[4]

The today organization of the societal function of feeding is considered unsustainable because of several reasons.

Multiple negative environmental impacts can be listed: the consumption of natural resources (land, water, non-renewable energy sources); the reduction of biodiversity; the perturbation of the global cycles of nytrogen, phosporus and CO_2; the generation of harmful outputs (soil, water and air pollutants) and waste; the limited attention to animal welfare and the recurrent generation of epizootic diseases (mostly originated in highly intensive and industrialized agricultural contexts). Some of these environmental impacts also feature social (health) side-effects; for example: chemical inputs can be harmful for farmers; local air pollutants (such as particulate matter) generated by food transportation contribute to higher morbidity and mortality rates (especially in urban areas); epizootic disease may affect humans too (as in the case of the 'mad cow').

Also social and economic negative impacts are relevant. First of all, the global agribusiness system is considered unequal for two main reasons: (1) because of asymmetric trade conditions of agriculture commodities between the global North (were agriculture is largely subsidized) and South (whose product importation in the North is often limited by tariffs), and (2) because of the (abuse of their) dominant position of global agribusiness players in the agriculture of developing and emerging economies. The latter mostly refers: to the limited access to land, also through the emergent phenomenon of 'land grabbing' (De Schutter 2011a); to the provision of high-priced inputs for agriculture; to low prices and salaries paid to farmers and agricultural workers; to bad working conditions; to low positive impacts on local economies. Then, the global agribusiness system is considered inefficient because of the apparent nonsense of the two twin phenomena of: (a) food scarcity and food waste and losses; (b) obesity and malnutrition. Finally, the ever increasing diffusion of highly transformed and standardized food products is

[3]The overall environmental impact of biofuels—with the only exception of sugar cane ethanol—is highly disputed (Eisentraut 2010).

[4]This paragraph largely draws on Sillig and Marletto (2013).

accused of negative effects on human health and of the reduction of both food cultural diversity and organoleptic quality.

Also because of the multidimensional negative impacts generated by the societal function of agribusiness, the overall attention to increase its sustainability is declined in several (and partially autonomous) discourses:

- 'Food for all' (or 'Modernization'). This is the today prevailing interpretation of the agribusiness overall aim: to deliver low-price, varied, easy-to-buy, safe and enjoyable food products to world consumers. Such an aim is ensured by highly productive agricultural techniques and complex global supply chains. Trademarks, brands and logos of both large-scale retailers and global producers of mass food products are worldwide known symbols of this discourse;
- 'Healthy food' (or 'Healthism'). Actually this is a patchwork of discourses resulting from the attention to multiple issues: the integrity and preservation of food products, the absence of toxic or harmful ingredients, the increasing interest for 'natural' food and 'good' nutrition, etc. This is why it is supported and implemented by—for example—global brands of mass food products, specialized producers of dietetic food, nutritional and functional products, and anonymous producers of very local products. Food traceability is an increasingly sensitive issue of this discourse;
- 'Environment-friendly food' (or 'Environmentalism'). This is part of the overall discourse on sustainable development and applies to several constituents of food systems in order to reduce the environmental impacts generated by the provision food. 'Green', organic (and also non-GMO) certifications are largely recognized as the visible part of this discourse;
- 'Food as local culture and local development' (or 'Localism'). In this discourse, food is considered as a constituent of local traditional knowledge that must be protected and revitalized, also in order to halt and reverse the trend towards the reduction of food cultural diversity and organoleptic quality. This discourse is incorporated in formal norms (such as those regulating geographical indications) and economic activities (such as local fairs and food and wine tourism initiatives). In more general terms, it can be considered as part of a wider discourse on local development that refers to both developed and developing countries;
- 'Food that is produced in a fair way' (or 'Equity'). This discourse focuses on the equity of the international trade of food products and is mostly associated to the 'fair trade' issue and to its organizations and certifications. It refers to the economic—and, in more general terms, working—conditions of marginalized producers and workers, in particular of those living in the global South.

Some of the above discourses partially overlaps. In particular: the organic sub discourse is shared by both the 'environmentalism' and 'healthism' discourses, and more and more by the 'localism' and 'equity' discourses too (the same applies to the non-GMO sub-discourse); the 'localism' discourse integrates from the 'environmentalism' discourse the reduction of transport intensity which results from shorter supply chains (or 'food miles'),[5] and from the 'equity' discourse the attention to both fair incomes to producers and fair prices to consumers. Also some trade-offs between discourses must be stressed. In particular, the 'equity' discourse is not consistent with the 'localism' and the 'environmentalism' discourses: because it is based on the import of food products from the global South (instead of the consumption of local products), and because more negative environmental impacts are generated by the needed intercontinental supply chains. The latter applies to the 'modernization' discourse too, because of its favor for free international trade of agricultural commodities and food products. The relation between the 'modernization' and 'healthism' discourses is two-sided: on one side, they share the attention to safe food, and also—but only with reference to dietary and nutritional products—to healthy food; on the other side they are inconsistent, as a diet mostly based on highly transformed food products it is not considered healthy (not to mention the campaigns against 'junk food', in particular as a component of the daily diet of the poorest) (Lawrence 2004).

3.2.3 The Socio-technical Map of Today's Food

All the above information can be represented with a ST-map that will be used to build future scenarios too. All following maps are based on two dimensions that represent the relevant knowledge leveraged by actors to foster innovation:

- Along the horizontal dimension all the political discourse on sustainable food described above are listed. The positioning of food systems with reference to this dimension reflects how their supporting networks of innovators interpret the current and future sustainability of food;
- Along the vertical dimension food supply chains are ranked from simple (bottom) to complex (top). Complexity of supply chains depends on both the distance covered and the number of steps; the latter in turn depends on the quantity of inputs used in both agriculture and industrial transformation, and on the level of industrial transformation. The positioning of ST systems with reference to this dimension reflects which kind of supply chains they usually manage.

In Fig. 3.1 all food systems and niches considered above are represented. Slowfood/Terra madre is the only individual actor represented because of its potential for innovation.

Worth stressing that all systems—except local ones—are able to manage very complex supply chains. Moreover, the ST-map of food highlights that most systems

[5]Actually, the scientific debate on "food miles" stresses that—thanks to a more efficient logistics (bigger vehicles with higher load factors)—longer supply chains can be less environmental harmful than shortes ones. In more general terms it is acknowledged that the environmental impact of food supply chains is product- and site-specific (Garnett 2003).

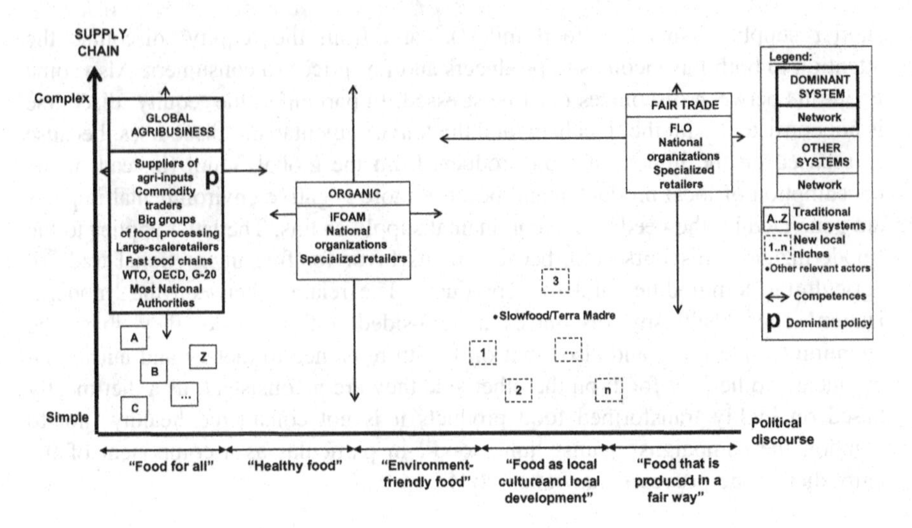

Fig. 3.1 The socio-technical map of feeding: current situation

cover more than one political discourse: the organic system hybridizes the discourses on human health and environmental sustainability; the global agribusiness system integrate into its discourse also the issue of healthy food (mostly interpreted in terms of safe food); some new local niches interpret the sustainability of food also in terms of environmental sustainability and equity; fair trade is extending its discourse from equity to local culture and environmental sustainability. Only most local systems and niches are centered on a single political discourse.

The currently dominant food policy is positioned inside the global agribusiness system in order to show that this system has a stable influence on global and national policies.

3.3 Sustainable Food Tomorrow: Biotech, Elitist or Organic?

3.3.1 The Potential for Change

The future of food will be conditioned by ongoing innovations in technologies, commercial practices and policies for sustainable food.

Two—maybe the most important—streams of technological innovations that should lead to sustainable food refer to the agricultural component of food systems.

The first stream of innovations is based on two new generations of applications of bio-technologies to agribusiness: while the first generation focused on the genetic modification of crops in order to make the use of chemical inputs more efficient, the second and the third generation aim at transforming the output. In particular: the second generation of GMOs is able to deliver products that reduce energy, storage and transport costs along the whole supply chain; the third generation should ease the transition from functional food to "nutraceuticals", that is, from products with added extra-nutrients to products where extra-nutrients are genetically integrated into its basic ingredients (The Economist 2011).

The second stream of innovations is based on the consideration of agriculture as part of the ecosystem, with a specific attention to the reproduction of ecosystem services that are provided to and by agriculture. This approach is named in different—and somehow confusing—ways; for example: "ecological agriculture", "sustainable intensification", "agricultural sustainability" (Tilman et al. 2002; Pretty 2008; Foley et al. 2011; Tscharntke et al. 2012). Through significant increases in nitrogen-, phosphorus- and water-use efficiency and the diffusion of ecological practices (such as, crop rotation and diversity, fallow periods, the use of composted animal waste as fertilizer, etc.) negative impacts of agriculture should be reduced, while its productivity should increase. Also the reduction of non-food productions (such as livestock feeds and bio-energy crops) can be considered. Some scholars suggest that this approach can be technology-neutral, thus there is no need "to be locked into a single approach whether it be conventional agriculture, genetic modification or organic farming" (Foley et al. 2011: 341).

Other, less important, commercial innovations for sustainable food must be taken into consideration: the selling of local, organic and 'fair trade' food products through large-scale retailers—mostly in developed countries; the use of ICT to increase the transparency of food products (e.g., through Apps that provide the consumer with supplementary information on food products, their ingredients and their producers) and gives place to "smart eating" (Choi and Graham 2014; Davies 2014); the application of nanotechnologies to different components of food systems: packaging, supply chain management, agricultural production (Lu and Bowles 2013; Sekhon 2014); the use of horticulture, gardening and cooking as educational or therapeutic practices (Davies 2014). Worth refers also to the diffusion of new generations of biofuels which should reduce the negative impacts on food security generated until now. In particular, second and third generation biofuels are not based on food crops and do not use land, respectively (Eisentraut 2010; Demirbas 2011, Applied energy).

Also some changes in the political approach to food must be considered as relevant to envisage future scenarios. In particular, two events have to some extent changed the global policy approach to the sustainability of food: (1) the food crises generated by 2007–2008 and 2010–2011 price spikes of agricultural commodities; and (2) the increasing knowledge and awareness about the bilateral relation between agriculture and climate change (Houghton et al. 1990; Godfray et al. 2011; FAO 2015; IPCC 2015). All the above has resulted in two relevant changes in global agriculture and food policy. First of all, the political issues of food security and sustainable agriculture are nowaday shared—even if with different approaches—by all global/transnational institutions.[6] This is true not only for those institutions that traditionally put food security as a top priority (UN 2015), but also for those institutions that were traditionally pro-market (World Bank 2015; Clapp 2015). Second, but not less relevant, some other specific issues and specific tools—that were abandoned because of the dominant pro-liberal approach to global agriculture and food policy—are now considered again in policy debates and implementations. In particular: the role of small producers for sustainable agriculture and food security is increasingly acknowledged by almost all global institutions (FAO 2011; World Bank 2011); the overall issue of self-sufficiency and the specific proposal of regional food reserves were center stage in the debate following the price crises (De Schutter 2011b; FAO et al. 2011; G20 Agriculture Ministers 2011; NEPAD 2011).

Also at the federal/national level a shift from socioeconomic to environmental issues can be registered. For example: since 2003 the EU common agriculture policy (CAP) is based on sustainability goals and standards, and the reformed 2014–2020 CAP will invest more than 100 billion Euros to support ecological agricultural practices (European Commission 2013).

3.3.2 Transition Pathways and Future Scenarios

Combining the ST-map of the current situations with on-going innovations, three transition pathways to future scenarios of sustainable food can be considered.

3.3.2.1 Transition Pathway to Scenario 1—'Biotech'

The 'Biotech' scenario (Fig. 3.2) results from a two-fold transition pathways involving the already dominant global agribusiness system. Also in order to respond to the increasing policy pressure for sustainable agriculture and food security, agricultural practices based on new generations of bio-technologies and nano-technologies diffuse worldwide, covering all the global South (were traditional local systems virtually disappear) and most of the global North (were new local systems and niches continue to reproduce). Such a transformation of agriculture is mostly backed by suppliers of biotechnological inputs that become more and more important into (and possibly lead) the supporting network of the global agribusiness system. As a result also the new generation of nutriceutical food products reaches commercial maturity.

At the same time, the retailing component of the system will be changed by a reconfiguration pathway: also in order to respond to the increasing demand of consumers for healthy, environment-friendly and fair food products, the commercial activity of the organic and fairtrade systems (and of many other new local

[6]For a critical view on this point see Wise and Murphy 2012.

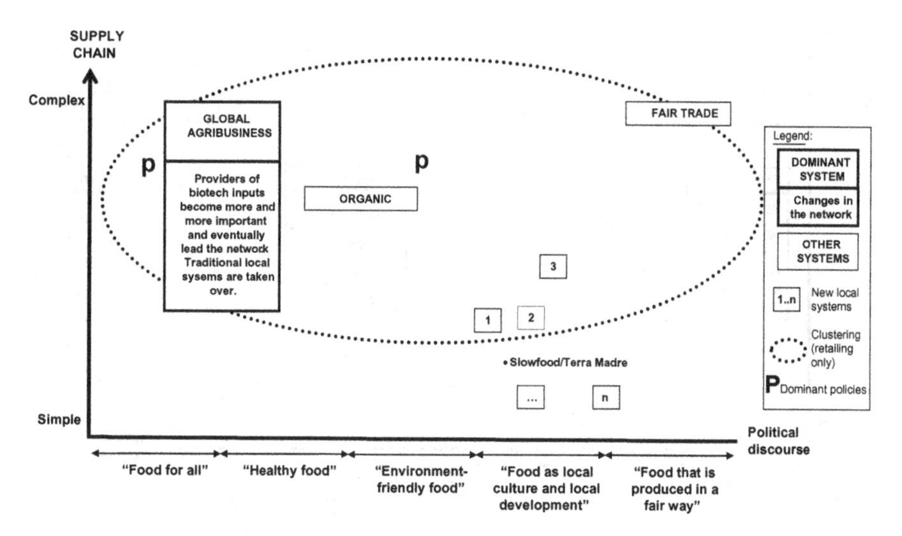

Fig. 3.2 The socio-technical map of feeding: 2030 'Biotech' scenario

systems and niches) will be partially integrated into the supply chains of large-scale retailers. Worth stressing that organic, fairtrade and most new local systems survive to this transition pathway by keeping an autonomous ability to reproduce and diffuse (mostly in the global North).

Both the techno-economic and sociopolitical dimensions of such a transition are mostly based on continuity. Supply chains remain complex and cover the whole world. Large-scale retailing becomes the hub of virtually all food products for virtually all humans. While higher productivity and free trade remain the main references of global/national policies for food security—and continue to counter-balance the pressure of policies for sustainable agriculture—, the global agribusiness system is able to gradually integrate the issues of healthy and environment-friendly food into its discourse. Such a continuity is possible because of the ability of the global agribusiness system to keep empowering also by getting support from policies implemented at both the global and the national level. Notwithstanding its dominant position, the sustainability of global agribusiness—even if transformed and reconfigurated—remains contended.

3.3.2.2 Transition Pathway to Scenario 2—'Elite Versus Mass'

The 'Elite Versus Mass' scenario (Fig. 3.3) features two dominant systems: the 'Elite food' and the 'Mass food'. The latter results from the transformation of the global agribusiness system that targets the lower segments of the global market, also by taking over most of the agribusiness activities of the global South (including many traditional local systems). Unlike what happens in the 'Biotech' scenario—the transformed global agribusiness system is not able to dominate the global North

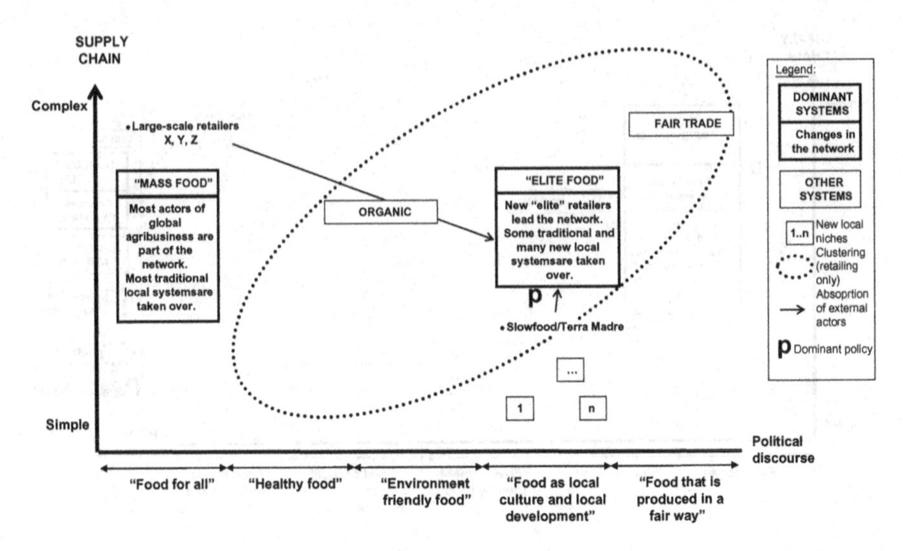

Fig. 3.3 The socio-technical map of feeding: 2030 'Elite Versus Mass' scenario

too. In this part of the world, a new network led by elitist large-scale retailers is able to serve the increasing demand for healthy, environment-friendly and local food; such an aim is reached by clustering the commercial activities of the organic and fair trade systems around the elitist retailers and by taking over many (both traditional and new) local systems. The structuring and networking activity performed by Slowfood/Terra Madre will play a relevant role in such a process of stable integration of local systems and niches from all over the world. At the same time, some new local systems continue to reproduce as autonomous niches. Also top market segments of the global South are served by the 'Elite food' system.

It is difficult to say if this scenario will result from a de-alignment and re-alignment transition pathway, or from a substitution. In the former case, traditional large-scale retailers spin-off from within the global agribusiness system and coalesce with other systems in order to sell top-quality food products. In the latter case a substitution takes place: the new elitist retailers empower from outside the global agribusiness systems, take over part of its retailing component and radically change it.

The 'Elite food' system is centred around the political discourse of localism, and its reproduction and empowerment is supported by national and regional policies for the protection of cultural diversity, taking place in the global North. But only if local development and small-scale producers become the main reference of global policies for sustainable agriculture and for the food self-sufficiency of the global South, then the 'Mass food' will be further destabilized and the 'Elite food' will be the only dominant system.

This scenario is less likely of the 'Biotech' just because two twinned process of radical change are needed: the destabilization of the today dominant global agribusiness system, and the emergence of a new approach to food policy.

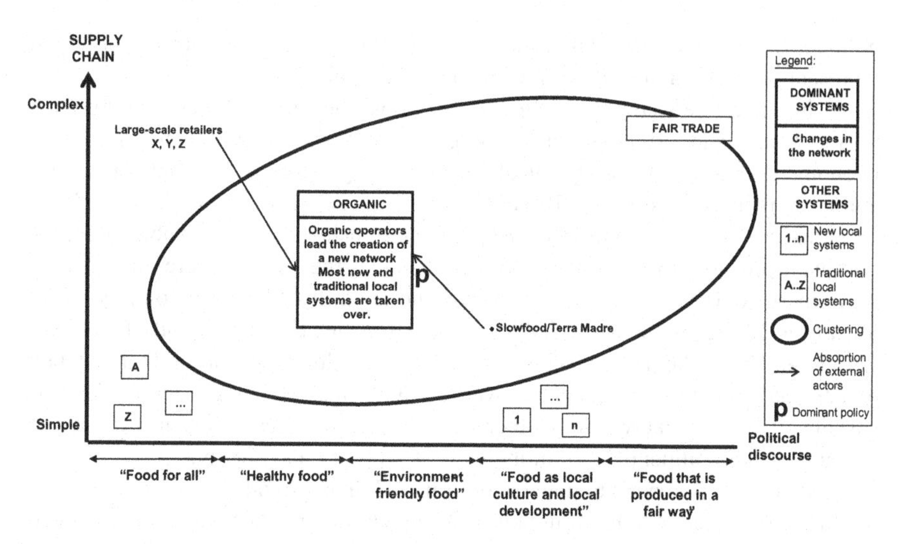

Fig. 3.4 The socio-technical map of feeding: 2030 'Organic' scenario

3.3.2.3 Transition Pathway to Scenario 3—'Organic'

In this case a de-alignment and re-alignment transition pathway gives place to a new dominant system led by organic operators who are able to integrate into a new network all the productive and commercial actors who are interested in converting to organic practices and to organic food products. In particular, the fair trade system and most traditional and new local systems cluster with the organic system. Moreover, also some large-scale retailers coming from the global agribusiness system join the organic network (Fig. 3.4).

The organic system keeps focusing on its hybrid discourse that integrates elements of both the environmental and health issues. Its gradual strengthening is eased by global, national and local policies that refer to organic practices as key for both sustainable agriculture, and the food security and self-sufficiency of the global South. Just because of these policies the organic system is able to take over the whole world societal function of feeding.

3.4 Final Remarks

This case study was aimed at understanding if the ST-map may prove valid to represent the ST dynamics of feeding, a societal function that features multiple political approaches to sustainability.

Notwithstanding the techno-economic complexities and multi-scale dynamics that characterize the provision of food, ST-maps of the current situation and of three alternative scenarios were successfully implemented. Both technological and political innovations were taken into consideration, without losing the ability to represent changes taking place in different geographical areas.

As in other case studies presented in this book, the strict link between dominant positions and supporting policies emerges as the main determinant of lock-ins phenomena that hinder those radical societal changes that are needed for sustainability transitions (SusTrans). 'Biotech' and 'Mass Versus Elite' are the more likely scenario just because in both cases the global agribusiness systems is able to keep its dominant position. More sustainable scenarios—as the 'Organic' presented here—may emerge only if new approaches to food policy diffuse, both horizontally—from local to local niches—and vertically from local niches to national and global institutions, and back. Actually, it is just the existence of global institutions that may help to generate top–down political innovations, thus making radical innovations for sustainability more likely than in other societal functions were global institutions are missing (or less influential), as in the case of mobility and energy.

Further work is needed to understand if the ST-map provided new knowledge on the future of sustainable food; in particular, by comparing the results provided here with those of other studies.

References

Al-Riffai P, Dimaranan B, Laborde D (2010) European Union and United States Biofuel Mandates. Impacts on World Markets. Technical Notes No. IDB-TN-191. Inter-American Development Bank. http://idbdocs.iadb.org/wsdocs/getdocument.aspx?docnum=35529623. Accesssed 26 May 2016.

Anderson K, Rausser G, Swinnen J (2013) Political economy of public policies: insights from distortions to agricultural and food markets. J Econ Lit 51:423–427

Bellemare MF, Carnes N (2015) Why do members of congress support agricultural protection? Food Policy 50:20–34

Brooks J (2014) Policy coherence and food security: the effects of OECD countries' agricultural policies. Food Policy 44:88–94

Choi JH, Graham M (2014) Urban food futures: ICTs and opportunities. Futur 62:151–154

Clapp J (2015) Food security and international trade–Unpacking disputed narratives. Background paper prepared for The State of Agricultural Commodity Markets 2015–16. FAO: Rome. http://www.fao.org/documents/card/en/c/63ba0e7a-c737-493a-942e-0a7d81ef87f3/. Accessed 26 May 2016

Clapp J, Fuchs D (2009) Corporate power in global agrifood governance. The MIT Press, Cambridge (MA) and London

Clive J (2014) Global Status of Commercialized Biotech/GM Crops: 2014. ISAAA Brief No. 49. ISAAA: Ithaca (NY). http://www.isaaa.org/resources/publications/briefs/46/default.asp. Accessed 26 May 2016

Davies AR (2014) Co-creating sustainable eating futures: technology, ICT and citizen–consumer ambivalence. Futures 62:181–194

Demirbas MF (2011) Biofuels from algae for sustainable development. Appl Energy 88: 3473–3480

De Schutter O (2011a) The green rush: the global race for farmland and the rights of land users. Harvard Int Law J 52:505–559

De Schutter O (2011b) G20 Action Plan addresses the symptoms, not the causes of the problem. UN Special Rapporteur on the Right to Food. http://www.srfood.org/en/g20-action-plan-addresses-the-symptoms-not-the-causes-of-the-problem. Accessed 08 Feb 2016

Eisentraut A (2010) Sustainable production of second-generation biofuels. Information Paper 2010 February. OECD/IEA: Paris. https://www.iea.org/publications/freepublications/publication/sustainable-production-of-second-generation-biofuels.html. Accessed 26 May 2016

European Commission (2013) The common agricultural policy (CAP) and agriculture in Europe–Frequently asked questions. http://europa.eu/rapid/press-release_MEMO-13-631_en.htm. Accessed 08 Feb 2016

FAO—Food and Agriculture Organization of the United Nations (2011) Save and Grow—A policymaker's guide to the sustainable intensification of smallholder crop production. FAO: Rome. http://www.fao.org/ag/save-and-grow/en/index.html. Accessed 26 May 2016

FAO–Food and Agriculture Organization of the United Nations (2015) Climate change and food systems: global assessments and implications for food security and trade. FAO: Rome (I). http://www.fao.org/documents/card/en/c/2d309fca-89be-481f-859e-72b27a3ea5dc/. Accessed 16 May 2016

FAO—Food and Agriculture Organization of the United Nations et al (2011) Price volatility in food and agricultural markets: policy responses. http://www.oecd.org/agriculture/pricevolatilityinfoodandagriculturalmarketspolicyresponses.htm. Accessed 26 May 2016

FiBL and IFOAM (2016) The world of agriculture farming–Statistics & emerging trends 2016. Research Institute of Organic Agriculture (FiBL) and IFOAM—Organics International: Frick (CH) and Bonn (D). http://www.organic-world.net/yearbook/yearbook-2016/pdf.html. Accessed 16 May 2016

Foley JA et al. (2011) Solutions for a cultivated planet. Nature 478:337–342

Garnett T (2003) Wise moves—Exploring the relationship between food, transport and CO_2. Transport 2000 Trust: London. http://www.fcrn.org.uk/research-library/wise-moves-exploring-relationship-between-food-transport-and-co2. Accessed 16 May 2016

Goodman D, Dupuis EM, Goodman MK (2012) Alternative food networks-knowledge, practice and politics. Routledge, Abingdon

G20 Agriculture Ministers (2011) Action Plan on Food Price Volatility and Agriculture. Ministerial Declaration, Meeting of G20 Agriculture Ministers: Paris, 22 and 23 June 2011. https://www.oecd.org/site/agrfcn/48479226.pdf. Accessed 16 May 2016

Godfray HCJ, Pretty J, Thomas SM, Warham EJ, Beddington JR (2011) Linking policy on climate and food. Science 331:1013–1014

Green K, Harvey M, McMeekin A (2003) Transformations in food consumption and production systems. J Environ Plann Policy Manage 5:145–163

Houghton JT, Jenkins GJ, Ephraums JJ (eds) (1990) Climate change: the IPCC Scientific Assessment-Report prepared for Intergovernmental Panel on Climate Change by Working Group I. Cambridge University Press, Cambridge

IPCC—Intergovernmental Panel on Climate Change (2015) Climate Change 2014: Synthesis Report. IPCC: Geneva (CH). http://www.ipcc.ch/report/ar5/syr/. Accessed 16 May 2016

Lawrence F (2004) Not on the label. What really goes into the food on your plate. Penguin Books, London

Lu J, Bowles M (2013) How will nanotechnology affect agricultural supply chains? Int Food Agribusiness Manag Rev 16:21–42

Murphy S (2006) Concentrated market power and agricultural trade. EcoFair Trade Dialogue Discussion Paper, No. 1. Heinrich Boll Foundation: Berlin (D). www.iatp.org/files/451_2_89014.pdf. Accessed 16 May 2016

NEPAD–New Partnership for Africa's Development (2011) AU/NEPAD Declaration on the G20 action plan on food price volatility and agriculture. http://www.nepad.org/ceo039s-office/news/2337/aunepad-declaration-about-g20-action-plan-food-price-volatility-and-agricul. Accessed 07 Feb 2016

Patel R (2003) Stuffed and starved. Portobello Books, London

Pretty J (2008) Agricultural sustainability: concepts, principles and evidence. Phil Trans R Soc B 363:447–465

Randall Wray L (2008) The Commodities Market Bubble–Money Manager Capitalism and the Financialization of Commodities. Public Policy Brief, No. 96/2008, The Levy Economics Institute of Bard College: Annandale-on-Hudson (NY). www.levyinstitute.org/pubs/ppb_96.pdf. Accessed 16 May 2016

Sekhon BS (2014) Nanotechnology in agri-food production: an overview. Nanotechnol Sci Appl 7:31–53

Sillig C, Marletto G (2013) La sostenibilità delle filiere agroalimentari–Valutazione degli impatti e inquadramento delle politiche (The sustainability of agrifood supply chains–Impact assessment and policy classification). Rapporti periodici 18. Isfort: Rome (I). www.isfort.it/sito/pubblicazioni/Rapporti%20periodici/RP_18_gennaio_2013.pdf. Accessed 16 May 2016

The Economist (2011) The 9 billion-people question—A special report on feeding the world. February 26th 2011, London. http://www.economist.com/node/18200618. Accessed 16 May 2016

Tilman D, Cassman KG, Matson PA, Naylor R, Polasky S (2002) Agricultural sustainability and intensive production practices. Nat 418:671–677

Tscharntke T et al. (2012) Global food security, biodiversity conservation and the future of agricultural intensification. Biol Conserv 151:53–59

UN–United Nations (2015) Transforming our world: the 2030 Agenda for Sustainable Development. Adopted by the United Nations General Assembly, 25 Sept 2015. http://www.un.org/ga/search/view_doc.asp?symbol=A/RES/70/1&Lang=E. Accessed 16 May 2016

Wise TA, Murphy S (2012) Resolving the Food Crisis—Assessing Global Policy Reforms Since 2007. IATP and GDAE: Medford. http://www.ase.tufts.edu/gdae/policy_research/resolving_food_crisis.html. Accessed 26 May 2016

World Bank (2011) Climate-Smart Agriculture– Increased Productivity and Food Security, Enhanced Resilience and Reduced Carbon Emissions for Sustainable Development. The World Bank: Washington. http://documents.worldbank.org/curated/en/2011/10/17486171/climate-smart-agriculture-increased-productivity-food-security-enhanced-resilience-reduced-carbon-emissions-sustainable-development. Accessed 26 May 2016

World Bank (2015) Future of Food. Shaping a Climate-Smart Global Food System. World Bank Group: Washington. http://documents.worldbank.org/curated/en/2015/10/25128378/future-food-shaping-climate-smart-global-food-system. Accessed 26 May 2016

WTO – World Trade Organization (1995) Ministerial Declaration. Adopted on 15 April 1994. https://www.wto.org/english/docs_e/legal_e/marrakesh_decl_e.htm. Accessed 16 May 2016

WTO—World Trade Organization (2001) Ministerial Declaration. Adopted on 14 Nov 2001. https://www.wto.org/english/thewto_e/minist_e/min01_e/mindecl_e.htm. Accessed 16 May 2016

Chapter 4
Multilevel Scenarios of Urban Mobility

Abstract In this chapter we show that the socio-technical map (ST-map) can be useful to represent sustainability transitions also when multi-scalar dynamics are at stake. This is the case of urban mobility: it takes place at a local level, but some of its constituents—actors, policies, technologies—are national or global. The analysis starts with the consideration of the ST systems and niches that concur to (and compete for) the provision of urban mobility: the dominant system of the individual car; the subaltern systems of public transport and the bicycle; the emerging car-sharing schemes. Also some locally dominant systems of integrated mobility are considered. These systems are then positioned into a ST-map built on two dimensions. The first dimension consider all the relevant political discourses on urban mobility: 'Mobility as a driver of development'; 'Mobility as a generator of harmful impacts'; 'Mobility as a determinant of the quality of urban spaces'; 'Mobility as a right'. The second dimension lists three alternative business models: 'sell vehicles', 'rent vehicles', 'manage transport systems'. Starting from the ST-map of the current situation of urban mobility—and from the consideration of the ongoing changes—three alternative scenarios are proposed. The 'Auto-city' scenario emerges from the reconfiguration of the existing 'individual car' dominant system and is generated by the absorption of the producers of batteries. In the 'Eco-city' scenario a coalition of urban networks supports a new political discourse of urban mobility and foster the creation of new urban systems of integrated mobility. In the 'Electri-city' scenario local and national electric operators takes over the individual car system because they are interested in the integration of smart grids and electric vehicles, also in order to increase grid stability and reduce demand-supply unbalances, in particular in the case of renewable sources. Multi-scalar dynamics are at the heart of the proposed scenarios of urban mobility: the 'Auto-city' and the 'Electri-city' scenarios mostly result from global dynamics, where niches are used for experimentation; the 'Eco-city' scenario emerges from a two-dimension diffusion process: horizontally, at the local level, where dominant positions "migrate" from an urban area to another; vertically, from the local to the national level, in order to gain greater political support.

G. Marletto et al., *Mapping Sustainability Transitions*,
SpringerBriefs in Business, DOI 10.1007/978-3-319-42274-9_4

Keywords Urban mobility · Car · Network of innovators · Sustainability transition · Socio-technical map · Scenario analysis

4.1 Introduction

Urban mobility is a multilevel societal function: it takes place at a local level, but some of its constituents are national or global: actors, policies, technologies, etc. As a result, transport systems that concur to (and compete for) the provision of urban mobility reproduce at different spatial levels, even simultaneously. Just to give two examples: the dynamics of the car system mostly occurs at a global level, but it must also confront national and local regulations that foster or hamper car use; public transport systems are usually rooted in local contexts, but they also build national and international associations in order to create and share a common culture, exchange information on best practices and innovations, and lobby.

The aim of this case-study is to understand if the socio-technical map (ST-map) can be useful to represent sustainability transitions also when multilevel dynamics of this kind are at stake. In particular, we will see if the ST-map is able to offer an overall representation of urban mobility that provides the reader with the 'big picture' of world innovations, without leaving in the shade local innovative practices. In the following paragraph all the elements of the ST-map of the current situation of urban mobility are explained, whereas 2030 alternative scenarios are built in Sect. 4.3.

4.2 Urban Mobility Today: The Dominance of the Car

4.2.1 Systems of Urban Mobility

The individual car is largely acknowledged as the dominant ST system of urban mobility, not only for its striking share of the mobility market (more than 80 % of total journeys in all developed countries, and an ever increasing modal share in emerging economies), as for the ability of its supporting network (where big global automotive and oil companies are the main core-actors) to influence institutions, policies and the society as a whole (Marletto 2011). This system is well centered on the propulsion technology of internal combustion, which powers 99 % of the currently circulating fleet. Toyota and Honda are an exception, as they have chosen the hybrid propulsion as a technological 'bridge' towards full electric propulsion: with millions of vehicles sold to date, hybrid cars are the only actual alternative to traditional cars (Hekkert and Van den Hoed 2006).

Public transport systems are usually considered as subaltern to the individual car system, because of low modal share (often less than 10 % of total mobility) and limited influence on national policies. With the only exceptions of urban areas where it is successfully integrated with other alternatives to the car (see below) public transport remains associated to the image of "transport for the poor" (Dennis and Urry 2009). The most relevant actors in the supporting networks of these systems are local: public transport companies, and urban and regional Authorities. At the urban and regional level these systems are usually able to obtain a significant amount of public resources which are used to build dedicated infrastructures and subsidize services. Since their birth, these systems have been able to manage the technology of electric propulsion and to plug-in vehicles (trolleys, tramways, trains, etc.) to the electric grid.

The bicycle is the other subaltern—if not marginal—system of urban mobility: in Northern America, Europe and Australia its average share of trips is negligible, that is, around 2 %. This figure is the result of a declining trend which started several decades ago in developed countries and more recently in emerging economies (where the use of bicycles is much more diffused, but rapidly declining). Notwithstanding these trends, more bicycles than cars are still produced worldwide: around 130 and 70 million per year, respectively; not surprisingly China is the larger producer and buyer of bicycles (including 13 million of electric bikes) (Sperling and Gordon 2009). Starting from the mid-70s the bicycle has experienced a revival supported by local and national coalitions of public actors and grassroots movements, both aiming at higher level of users' health, urban livability and environmental quality. In some countries these coalitions have been able to gain an influence on national policies too: in The Netherlands, Denmark and Germany more than 10 % of today mobility is assured by bicycles, and in some pro-bike cities of these countries bicycles serve more than 25 % of total trips. These cases resulted from a multilevel action, combining national plans and guidelines with the local provisions of cycling routes, dedicated parking and other supporting measures (traffic calming, intersection modifications, integration with public transport, training and education, etc.) (Pucher and Buehler 2008). Recent figures signal the increasing use of bicycles in some North American cities too (Portland, Minneapolis, Vancouver, etc.), with a resulting share which is still around 3–5 % of commuters, but reaches 6–8 % in central areas; these trends are mainly caused by the building of new bike lanes and pathways by local Administrations (Pucher et al. 2011).

Sharing schemes are a multitude of niches and systems which provide members with access to a vehicle for short-term use. A fleet of vehicles and specific technologies for the remote control of vehicles are the standard equipment of these systems; specific Apps to locate available vehicles are a more and more diffuse alternative to dedicated parking areas. Sharing has experienced a rapid extension from cars to bicycles, with the Parisian "Velib" bike-sharing scheme as the most relevant example. Worldwide the most recent figures count more than 1000 cities hosting a bike-sharing scheme (with around 1.25 million bicycles) and almost five millions carsharing[1] members (with more than 100,000 vehicles) (MetroBike 2016; Shaheen and Cohen 2016). It must be stressed that most of the pioneering

[1]In the UK carsharing schemes are known as 'car clubs' and carsharing is a synonymous of car pooling, i.e. the shared use of a car owned by one of the travelers.

experiences of carsharing were initially supported by non-profit actors (e.g., ShareCom in Switzerland—then merged in Mobility—and Cambio in Germany) and then evolved into commercial initiatives. Other established carsharing schemes are: 'Greenwheels' in the Netherlands and Germany, 'Zipcar' in the US and UK, 'Car2go' in Europe and Northern America. 'Autolib' in Paris is the most important case of carsharing with electric cars. Innovative managers of rental systems for both passenger and freight transport can be considered part of these niches too.

In several world urban areas successful local niches and systems of integrated mobility have generated a reduction of the use of individual cars down to 40 % of total mobility (or less). In this areas, all alternatives to the individual car—public transport, sharing schemes, "soft mobility" (that is, bicycles + pedestrians)—are integrated by hard and soft measures of urban planning and transport policy. Examples of already established systems of this kind can be found in some capital cities too, such as Amsterdam, Bogotà, Copenhagen, Paris, Stockholm.

4.2.2 Discourses on (Sustainable) Urban Mobility

Today urban mobility is considered unsustainable because of its negative environmental, social and economic impacts. Environmental impacts mostly refers to the generation of greenhouse gasses by internal combustion vehicles, in particular CO_2; even electric urban vehicles (such as trolleys and subways) generate this kind of side effect when the electricity they use is produced from nonrenewable sources. Most social impacts of urban mobility actually are health impacts, in particular: deaths and diseases caused by air pollutants emitted by internal combustion vehicles (such as particulate matter) and by road accidents (of which a relevant share involves pedestrians and cyclists). Other social impacts refer to the transformations of urban space induced by road transport; in particular: the reduction of space that is available for other uses than mobility, and the fragmentation of urban space caused by transport infrastructures. The most important negative economic impact of urban mobility is the waste of time caused by the congestion of urban roads and the saturation of parking areas.

It is apparent that most of the above negative impacts are associated to the use of cars as an individual mean of urban mobility; this is why the political debate on the sustainability of urban mobility is mostly centered on the (excessive) use of cars in urban areas. And it is just the car that stays center stage of one of the political discourses on urban mobility: 'Mobility as a driver of development' (or 'Modernization'). Indeed, the car is considered as a driver of positive impacts: because of the huge investments and jobs it brings along, and because it bestows to all individuals the privilege of free circulation that used to be the privilege of the rich. But not only the car can be a generator of these positive impacts: also other public works for urban mobility (such as, infrastructures for urban railways, metros and trolley cars) generates investments and jobs, and—just because of the congestion generated by the diffusion of cars—other transport systems may provide the users with actual free circulation.

Also another political discourse on urban mobility mostly focusses on the car, in particular on its negative impacts: 'Mobility as a generator of harmful impacts' (or 'Sustainability'). In this case the attention is towards all actions that can reduce the negative impacts generated by urban mobility, and in particular by the excessive use of individual (internal combustion) cars. Investments for public transport infrastructures and services, the re-organization of urban space in favor of pedestrians and cyclists, incentives to buy electric vehicles, promotion of sharing systems, access and parking charges: these are just some examples of the plethora of interventions that are discussed and implemented worldwide in order to reach higher levels of sustainable urban mobility.

A further political discourse on urban mobility actually derives from a wider debate on urban planning: 'Mobility as a determinant of the quality of urban spaces' (or 'Urban livability'). The organization of urban space has changed since motorization diffused in cities: what was freely accessible it has been strictly regulated; what was a living space, it has become a transport infrastructure. As a result the quality of urban spaces has worsened. This is why a worldwide movement reclaiming quality urban spaces also aims at limiting or banning motorized traffic—in particular in residential areas—and at promoting non-motorized mobility (e.g., through pedestrian areas, traffic calming measures, car-free neighborhoods, paths reserved to pedestrians and bicycles, etc.).

The last political discourse we consider here is also the oldest one: 'Mobility as a right' (or 'Welfarism'). Today the public provision and subsidization of collective transport services is diffused worldwide. This is the result of a process that started at the beginning of the last century in European and Northern America cities, where public transit systems were realized (in most cases through the nationalization and further development of privately managed lines and networks) in order to ensure to all citizens the right to mobility. These urban transport services were (and still are) considered as a constituent of the welfare state, and as such they have been involved in the more recent debate on the privatization and liberalization of public utilities, but with more limited changes than in other sectors (e.g., energy and telecom).

These discourses are not completely independent of each other. Because of the diffusion of individual cars in developed and emerging countries, the political discourse of 'Welfarism' has become less and less relevant: for only a marginal share of urban residents public transport is the only alternative to get around. Also because of this change, public transit is more and more viewed as a mean to aim at urban sustainability and livability. Moreover, the discourses of 'Sustainability' and 'Urban livability' feature a large overlapping, in particular because urban areas where motorized traffic is banned or limited, are also areas with reduced level of air pollution, accidents and noise. Also the discourses of 'Sustainability' and 'Development' partially overlaps, but only when actions for sustainable urban transport incorporate huge investments and jobs. This is not the case of actions for 'Urban livability' that are mostly realized through soft measures.

4.2.3 *The Socio-technical Map of Today's Urban Mobility*

All the elements that are considered relevant for the reproduction of the societal function of urban mobility are represented in the following ST-map (Fig. 4.1). The ST-map of urban mobility does not refer to a specific urban situation; on the contrary, an explicit attempt is made to deliver an analysis representing all the (both global and local) dynamics that are relevant at a global scale.

The ST-map of urban mobility is based on two dimensions:

- Discourses on urban mobility described above are listed along the horizontal dimension;
- Business models are listed along the vertical dimension. Three typologies are considered (from bottom to top): 'Sell vehicles'; 'Rent vehicles'; 'Manage transport systems'.

The positioning of ST systems and niches of urban mobility with respect to such two dimensions reflects the relevant knowledge that ST systems and niches of urban mobility leverage in order to foster technological, organizational and political innovations.

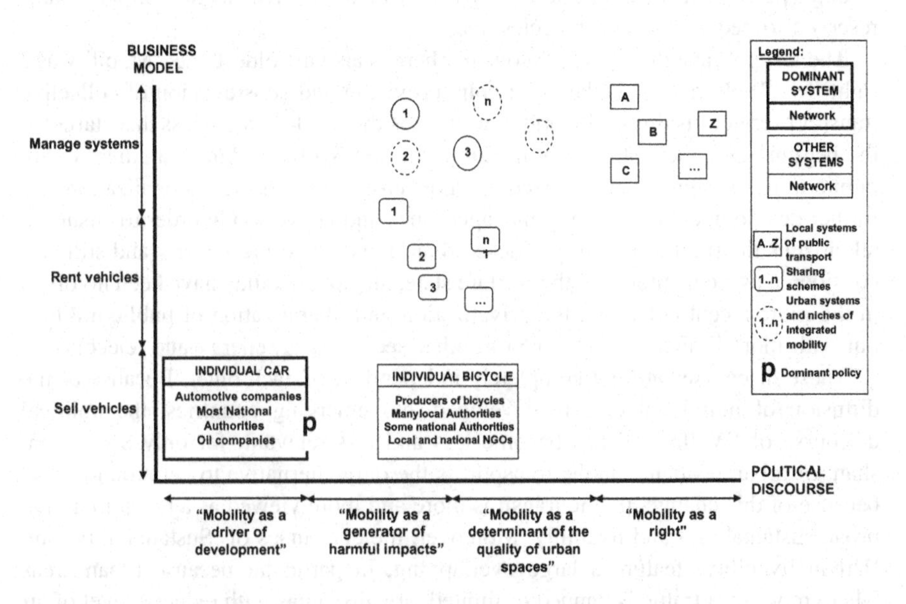

Fig. 4.1 The socio-technical map of urban mobility: current situation

4.3 The Future of Urban Mobility: Auto-City, Eco-City or Electri-City?

4.3.1 The Potential for Change

The societal function of urban mobility as a whole is under the pressure of a twofold quest for global sustainability and urban livability: more and more diffused limits to urban circulation and tighter and tighter emission standards on vehicles are just the two most relevant examples of an increasing and diffusing number of local, national and global policies that force all the relevant actors to change the way in which today's urban mobility is provided.

Starting from its dominant position, the 'individual car' system is already looking for an effective answer to such demand for change. Fiat-Chrysler and Volkswagen are just two examples of the more conservative—and until today, more diffused—innovation strategy, based on efficient internal combustion and down-sizing: a strategy implemented by most leading manufacturers too, such as Daimler, Ford, Hyundai, Nissan, Honda and Toyota (Wells et al. 2012). Toyota and Honda are also the main promoters of the "hybridization" of the car; they have chosen the hybrid propulsion as the entry-point to a process of technological innovation which, at the same time, is: (a) compatible with the current core competences, sunk investments, dominant design and interdependencies of the automotive industry, and (b) flexible enough to allow the future access to full electric cars (Avadikyan and Llerena 2010). Some other leading automotive companies—e.g. Citroen and Mitsubishi—jumped directly into the full electric car technology, but mostly as a residual option to internal combustion cars. On the contrary, this is the strategy implemented by most Chinese newcomers who are entering the technology of full electric propulsion without the sunk costs of previous investments. Also small specialized assemblers and manufacturers (as Heuliez, Pininfarina, Valmet, etc.) are trying to develop their EVs on a limited productive and commercial basis (Wells 2010). Maybe more important, producers of batteries—and other electric and electronic components—play a more and more relevant role in the trajectory of electrification (Orsato et al. 2012): some of them are implementing autonomous strategies, such as: Bolloré, who developed the Parisian "Autolib" carsharing scheme with Pininfarina (the Italian producer of the electric "Bluecar") and recently signed a letter of intent with Renault to collaborate on carsharing operations and eventually develop a new small electric vehicle; BYD (Build Your Dreams), a private Chinese producer of batteries for computers and cellular phones, who is now producing cars. Some other car producers are trying—at very different scale of testing and marketing—to integrate some elements of the 'rent vehicles' and 'manage systems' business models into the car system: Nissan-Renault already launched the mixed option of selling full-electric cars and renting batteries; Daimler (with its electric Mini) and BMW (with its electric Smart) are promoting two vehicle-to-grid (V2G) tests, in cooperation with two energy suppliers: the Italian Enel and the Swedish Vattenfall, respectively (OECD et al. 2014). Moreover, an increasing number of electric utilities is involved in partnerships related to the diffusion of EVs. But the most promising approach is the stable integration into big automotive companies of those actors that may provide the needed new competences, in particular the more innovative producers of batteries. This kind of transition may be hampered by the delegitimization of the supporting coalition of the 'individual car'—and of their political discourse—and by the consequent emergence of policies explicitly aimed at reducing car use and ownership. But this event is highly unlikely, because of the current global dominant position of the supporting coalition of the 'individual car' system, and the resulting ability to keep gaining a political support for incentive schemes to "green" (or "greenwash"?) the car.

Another relevant endogenous dynamic refers to the increasing number of cities where a transition to a new system of integrated mobility is gaining ground, or even—in the most dynamic areas—is already accomplished. This is mainly the result of the ability of local networks of innovators to influence the urban policy arena, by fostering a new political discourse that hybridize sustainability and urban livability issues, and by obtaining a radical change in actual policies. A not secondary constituent of this political process is the involvement of public transport that moves away from the political discourse of 'mobility as a right'. In some cases (e.g., Netherlands, Switzerland) these innovators has gained political legitimacy at the national level too.

The last—but highly relevant—potential for change is coming from some very powerful outsiders that are currently involved only in some niche experiments. First of all, the electric operators (producers of electricity and managers of electric grids). These actors already feature high level of competences, resources and legitimacy; in particular, they are able to found their actions on a successful hybridization of the political discourses of 'modernization' and 'sustainability'. Their massive entry into the societal function of urban mobility may therefore have a dramatic (and even disruptive) effect on all today's actors and systems. This is why the enlistment of actors and systems of urban mobility in a new supporting coalition led by electric operators may finally result in a takeover of the current dominant position of the 'individual car'.

Also the actors involved in the development of a full self-driving car should be considered. Some of them—research bodies, suppliers of electronic components, etc.—are working in joint projects with automotive companies (e.g., Volvo, Daimler, GM, Audi, Nissan, etc.), but some other external (and powerful) actors are realizing autonomous initiative (with Google in a leading position) (Lari et al. 2015).

4.3.2 *Transition Pathways to Future Scenarios*

Specific transition pathways will be triggered and deployed if one or another of the above potential for changes will prevail; the three more likely alternatives are described in details in this section.

4.3.2.1 Transition Pathway to Scenario 1—'Auto-City'

This first transition pathway emerges from the reconfiguration of the existing 'individual car' dominant system and is generated by the absorption of new industrial actors, in particular producers of batteries. This extension of the coalition is aimed at acquiring crucial competences on the electric car; indeed, this technology is increasingly considered by the automotive industry as the long-term response to the gradual—but significant—shift of dominant policies from the political discourse of 'modernization' to those of 'sustainability' and 'urban livability' (Elzen et al. 2004; Dennis and Urry 2009). The integration into the car of other technologies will also continue, in particular to achieve higher levels of connection to the internet of things and self-driving.

As some analysts suggest, the battery may become the most important element in the electric car value chain (Zhou et al. 2011); consequently, producers of batteries may become 'core-actors' of this system. At the same time—because of the changing mix of energy sources used to power cars—oil companies should lose their position as a core-actor or eventually change their core-business, while managers of electric grids may evolve their role from mere suppliers of an essential utility to members of the coalition supporting the system.

Along the transition pathway the business model remains focused on selling cars to individual consumers, but—if also carsharing schemes are steadily integrated—it could be extended to the 'rent' option too. The share of electric cars steadily increases along the transition pathway and reaches the threshold of 35 % of car sales in 2030. Two different global phenomena can be detected: the first one in developed countries, where the rapid diffusion of hybrid cars is made possible by consumers and producers who gradually unlock from internal combustion; the second one in emerging economies, where the boom of full EVs is supported by newcomers—with new Chinese automotive companies playing a relevant role—who benefit from the lower barriers to entry associated with the technology of electric compared to internal combustion cars. Newly urbanized families in emerging economies contributes to support the demand side of these rising markets (Charles et al. 2011; Dijk et al. 2013). These industrial global trends are eased by national (e.g., China) and federal (e.g., EU, USA, India) policy schemes that provide incentives to car buyers and funds to builders of recharging infrastructure.

If one looks at a likely ending-point of this first transition pathway (see Fig. 4.2) the 'individual car' system keeps its dominant position on urban mobility; other systems of urban mobility—public transport, the individual bicycle—remain in a subaltern position. At the same time, it is under dispute if this transition pathway will reach the decarbonization targets set by an increasing number of legislations (Pasaoglu et al. 2012). At least three factors play against this possibility: (1) the too low rate of diffusion of electric cars; (2) the "rebound" effect on energy consumption that may be generated by an increasing amount of kilometers driven by cars; (3) the high-carbon energy-mix used to power electric cars in some countries, with China as a global worst-practice (Doucette and McCulloch 2011).

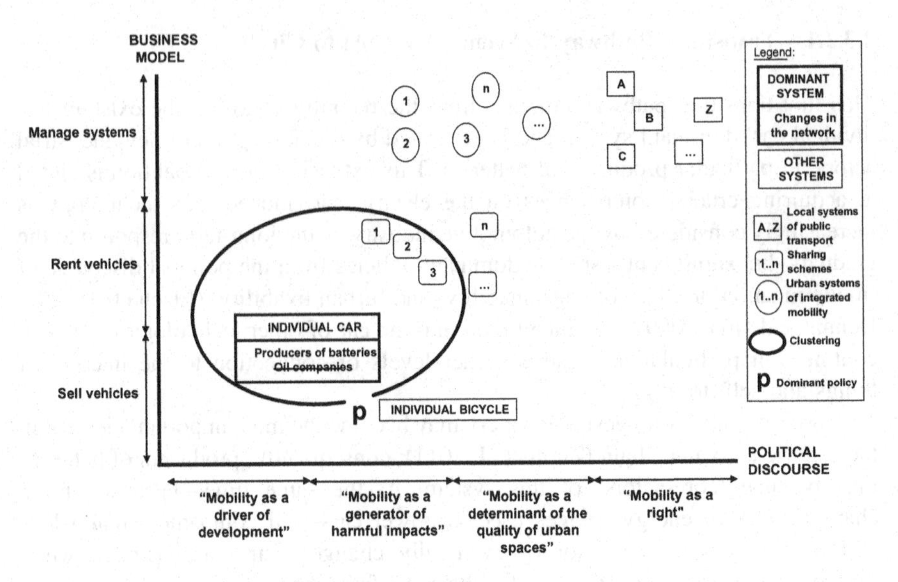

Fig. 4.2 The socio-technical map of urban mobility: 2030 'Auto-city' scenario

4.3.2.2 Transition Pathway to Scenario 2—'Eco-City'

In this transition pathway coalitions of urban networks support a new political discourse of urban mobility and foster the creation of new urban systems of integrated mobility (Vergragt and Brown 2007). Along the pathway the main transformative mechanism in place is the clustering—first locally and then nationally—of existing and emerging niches and systems of mobility. In particular: more and more local public transport systems move away from the political discourse of 'mobility as a right', and the individual bicycle system gradually moves from the 'sell' to 'rent' business model. Producers of EVs are gradually absorbed into the system, mostly as suppliers of all kind of vehicles for sharing schemes and fleet operators; providers of ICT devices for individual transport planning are absorbed too (Dijk et al. 2013). Moreover, Google—or other non-automotive producers of self-driving cars—may enter this new network as providers of new (fully floating) carsharing schemes.

The actual dynamics of the 'Eco-city' transition is the result of two parallel forces which must be analyzed with a spatial key: at the national level the gradual de-alignment of most institutional, economics and technological constituents of the 'individual car' system takes place, mostly because of a radical shift of the dominant policy approach towards 'urban livability'; at the same time, these and other constituents are gradually 're-aligned' into an increasing number of urban systems of integrated mobility.

Figure 4.3 represents the ending point of this transition. In 2030 stable national coalitions of local networks support the reproduction of urban systems of integrated mobility, while the individual car is in a subordinate position, supported by the few

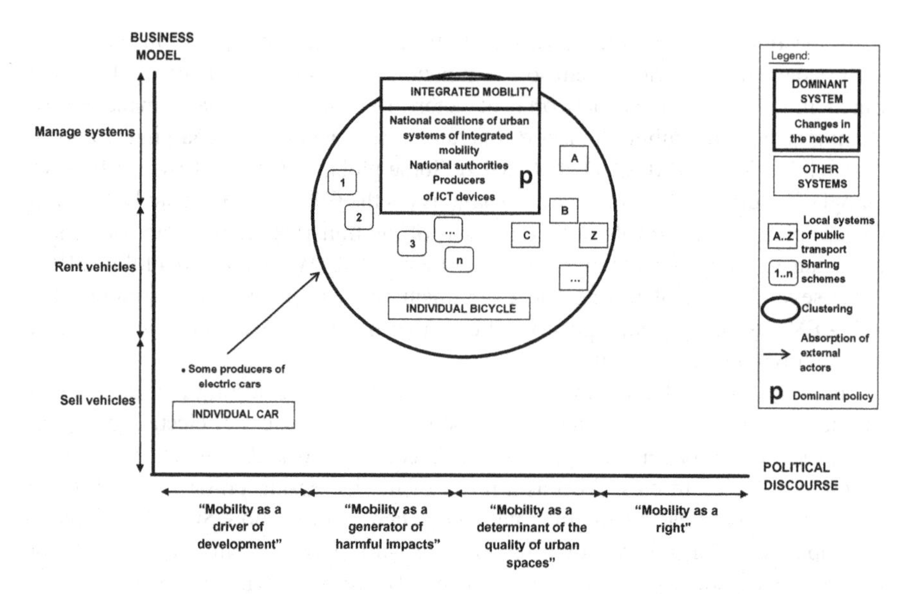

Fig. 4.3 The socio-technical map of urban mobility: 2030 'Eco-city' scenario

surviving world automotive companies. The 'Eco-city' scenario is more sustainable than the 'Auto-city' because the substitution of car use with non-motorized transport, shared vehicles and public transport—together with the diffusion of electric propulsion—not only can meet tight environmental targets without the need of an aggressive decarbonization of electric generation, but can also significantly increase urban livability (Bristow et al. 2008; McCollum and Yang 2009). But the 'Eco-city' scenario is less probable than the 'Auto-city' because its actual deployment depends on changes taking place at all level of social life: at the micro level, the spread of urban lifestyles which are no more based on the use of individual cars; at the meso (i.e., systemic) level, the creation and empowering of urban networks of innovators; at the macro level, the destabilization of the 'individual car' system and the diffusion of systems of integrated mobility from pioneering to laggard cities.

4.3.2.3 Transition Pathway to Scenario 3—'Electri-City'

In this transition pathway local and national electric operators are interested in the adaptation of their systems to the diffusion of EVs, because they aim at the new frontier of smart grids (SGs). SGs are able to exchange electricity with distributed energy resources, also in order to increase grid stability and reduce demand-supply unbalances, in particular in the case of renewable sources (Mullan et al. 2012).

The 'Electri-city' transition pathway can be divided into two phases: the first one for local testing (2013–2020) and the second one for consolidation at the global scale (2021–2030).

In the first phase, several experiments in niche cities take place, in particular: (a) to adapt the electric system to the function of mobility, and (b) to check the functioning of new local networks of actors (which may include: managers of sharing schemes, public transport operators, managers of recharging stations, research bodies, etc.) (OECD et al. 2014). Some global actors (as Tesla for EVs, or Google for self-driving cars) and associations (as the C40 Cities Climate Leadership Group) play a relevant role in this first part of the transition as promoter of tests at all urban scales, from medium towns to megacities (Wiederer and Philip 2010). After several years of testing and experimenting, it is increasingly apparent that SG + EV systems generate positive effect in terms of both economic development and environmental sustainability.

In the following decade, the positive results of previous testing fuel the interest of operators coming from different sectors: not only managers of electric grids, but also producers of batteries, suppliers of ICT components and—last but not least—producers of plug-in cars. Also as a result of the increasing pressures of all these operators on political institutions, national schemes to support SG + EV systems are implemented in several countries, starting with those featuring higher shares of electric generation from renewable sources (Denmark, Germany, Spain, France, etc.) (Leurent and Windisch 2011). Already established purchase subsidies are restricted to plug-in electric cars only and are integrated with investments on old and new infrastructures (e.g., metropolitan railway networks and SGs). Moreover, common standards on grids, plugs and batteries are introduced to further catalyze the diffusion of SG + EV systems (Egyedi and Spirco 2011).

The final scenario emerging from this transition pathway is represented in Fig. 4.4. This is the result of a successful "takeover bid" on the 'individual car'

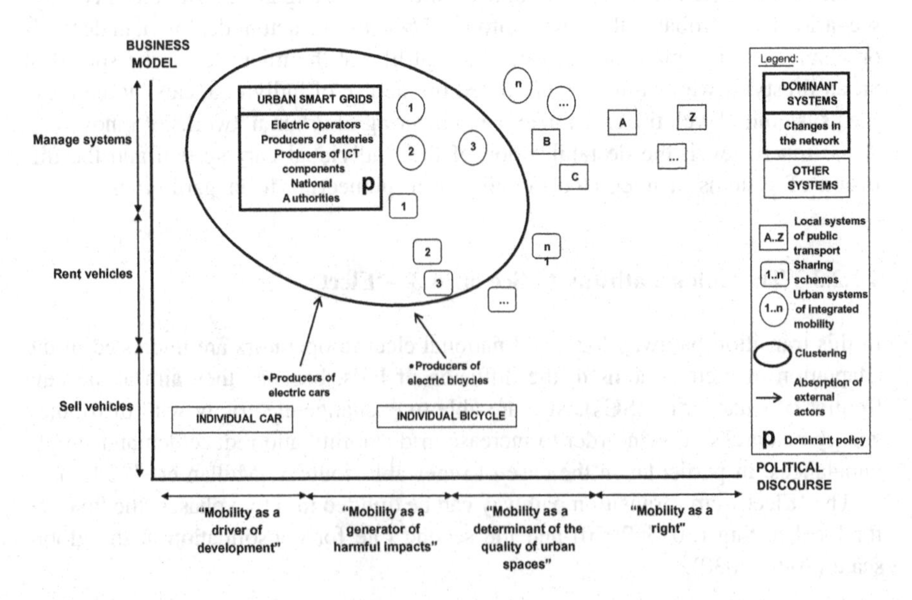

Fig. 4.4 The socio-technical map of urban mobility: 2030 'Electri-city' scenario

system which is launched by enactors (then core-actors) coming from another societal function. The environmental sustainability of this scenario is conditioned by the energy mix used to generate electricity and—what matters most—its likelihood is crucially conditioned by successful testing and the actual exploitation of latent economies of scale.

4.4 Final Remarks

The ST-map of urban mobility proved valid to generate alternative scenarios that represent the predictable effects of innovations taking place at multiple levels, from local to global. In particular, it helps to understand the impact on the sustainability transition (SusTran) of urban mobility of the co-evolution of innovation networks and supporting policies. And it is just this co-evolutionary process that may take place differently—in terms of both direction and speed—at the urban/local or national/global level. This is why in the ST-map of the current situation of urban mobility dominant positions co-exist: the individual car at the global level, and many systems of integrated mobility at the urban level. The explicit consideration of such a multilevel innovation dynamics is at the heart of the ST scenarios of urban mobility proposed here. The 'Auto-city' and the 'Electri-city' scenarios mostly result from a transition pathways that deploys at the global level (with local niches for testing and experimentation). The 'Eco-city' scenario emerges from a sequential process: first, transition pathways take place in several urban areas, thus generating new locally dominant positions; then, the latter diffuse horizontally (that is, from a local area to another) and vertically (that is, to the national/global level, in order to gain greater political support). The analysis of such a multi-scalar process must be realized in greater detail, also through specific case studies. We start, in the following chapter, with the analysis of the SusTran that resulted in the creation of the 'Eco-city' system of Freiburg (D).

References

Avadikyan A, Llerena P (2010) A real options reasoning approach to hybrid vehicle investments. Technol Forecast Soc Change 77:649–661

Bristow AL, Tight M, Pridmore A, May AD (2008) Developing pathways to low carbon land-based passenger transport in Great Britain by 2050. Energy Policy 36:3427–3435

Charles MB, To H, Gillet P, von der Heidt T, Kivits R (2011) Transport energy futures: exploring the geopolitical dimension. Futures 43:1142–1153

Dennis K, Urry J (2009) After the car. Polity Press, Cambridge

Dijk M, Orsato RJ, Kemp R (2013) The emergence of an electric mobility trajectory. Energy Policy 52:135–145

Doucette RT, McCulloch MD (2011) Modeling the CO_2 emissions from battery electric vehicles given the power generation mixes of different countries. Energy Policy 39:803–811

Egyedi T, Spirco J (2011) Standards in transitions: catalyzing infrastructure change. Futures 43:947–960

Elzen B, Geels FW, Hofman PS, Green K (2004) Socio-technical scenarios as a tool for transition policy: an example from the traffic and transport domain. In: Elzen B, Geels FW, Green K (eds) System innovation and the transition to sustainability. Edward Elgar, Cheltenham

Hekkert M, Van den Hoed R (2006) Competing technologies and the struggle towards a new dominant design: the emergence of the hybrid vehicle at the expense of the fuel-cell vehicle? In: Nieuwenhuis P, Vergragt P, Wells P (eds) The business of sustainable mobility: from vision to reality. Greenleaf Publishing, Sheffield

Lari A, Douma F, Onyiah I (2015) Self-driving vehicles and policy implications: current status of autonomous vehicle development and minnesota policy implications. Minn J Law Sci Technol: 735–769

Leurent F, Windisch E (2011) Triggering the development of electric mobility: a review of public policies. Eur Transp Res Rev 3:221–235

Marletto G (2011) Structure, agency and change in the car regime: a review of the literature. Eur Transp 47:71–88

McCollum D, Yang C (2009) Achieving deep reductions in US transport greenhouse gas emissions: scenario analysis and policy implications. Energy Policy 37:5580–5596

MetroBike (2016) The bike-sharing world—year end data 2015. The bike-sharing blog. http://bike-sharing.blogspot.it/2016/01/the-bike-sharing-world-year-end-data.html. Accessed 16 May 2016

Mullan J, Harries D, Braunl T, Whitely S (2012) The technical, economic and commercial viability of the vehicle-to-grid concept. Energy Policy 48:394–406

OECD, Rocky Mountain Institute, IEA (2014) EV city casebook, OECD, Paris. http://www.cleanenergyministerial.org/News/2014-ev-city-casebook-profiles-50-big-ideas-in-electric-mobility-448. Accessed 16 May 2016

Orsato DJ, Dijk M, Kemp R, Yarime M (2012) The electrification of automobility. In: Geels WF, Kemp R, Dudley G, Lyons G (eds) Automobility in transition? A Socio-technical analysis of sustainable transport. Routledge, Abingdon

Pasaoglu G, Honselaar M, Thiel C (2012) Potential vehicle fleet CO_2 reductions and cost implications for various vehicle technology deployment scenarios in Europe. Energy Policy 40:404–421

Pucher J, Buehler R (2008) Making cycling irresistible: lessons from the Netherlands, Denmark and Germany. Transp Rev 28:495–528

Pucher J, Buehler R, Seinen M (2011) Bicycling renaissance in North America? an update and re-appraisal of cycling trends and policies. Transp Res Part A 45:451–475

Shaheen S, Cohen A (2016) Innovative mobility carsharing outlook, winter 2016. Transportation Sustainability Research Centre, University of California, Berkeley. http://tsrc.berkeley.edu/sites/default/files/Innovative%20Mobility%20Industry%20Outlook_World%202016%20Final.pdf. Accessed 16 May 2016

Sperling D, Gordon G (2009) Two billion cars: driving toward sustainability. Oxford University Press, New York

Vergragt PJ, Brown H (2007) Sustainable mobility: from technological innovation to societal learning. J Clean Prod 15:1104–1115

Wells PE (2010) The automotive industry in an era of eco-austerity: creating an industry as if the planet mattered. Cheltenham, Edward Elgar

Wells PE, Nieuwenhuis P, Orsato DJ (2012) The nature and causes of inertia in the automotive industry. In: Geels WF, Kemp R, Dudley G, Lyons G (eds) Automobility in transition? a socio-technical analysis of sustainable transport. Routledge, Abingdon

Wiederer A, Philip R (2010) Policy options for electric vehicle charging infrastructure in C40 cities. Report for Stephen Crolius, Director—Transportation, Clinton Climate Initiative, Harvard Kennedy School. https://www.innovations.harvard.edu/sites/default/files/1108934.pdf. Accessed 16 May 2016

Zhou L, Watts JW, Sase M, Miyata A (2011) Charging ahead: battery electric vehicles and the transformation of an industry. Deloitte Rev 7. http://dupress.com/wp-content/uploads/2010/07/US_deloittereview_Charging_Ahead_Battery_Electric_Vehicles_Jul10.pdf. Accessed 16 May 2016

Chapter 5
Freiburg: From 'Auto-City' to 'City of Short Distances' (1945–2010)

Abstract In this chapter we analyse the sustainability transition pathway that made Freiburg an 'Eco-city' where the modal share of slow mobility is more than 50 % (pedestrians: 24 %; cycling: 28 %), and the car is less relevant (30 %). Through the sequential use of three ST-maps we are able to reconstruct the sustainability transition of Freiburg: from an emerging 'Auto-city', where private cars increase of importance (first ST-map: 1969), through a time of coexistence of four different systems (individual car, public transport, individual bicycle, walkability) (second ST-map: 1979), until the polarization in two systems: the individual car and the so-called 'City of Short Distances' (third ST-map: 2010). The reproduction of these two systems takes place at different scales: local actors support the 'City of Short Distances'—which is today dominant in Freiburg—while national and international actors keep supporting the individual car.

Keywords Freiburg · Car · Urban mobility · Network of innovators · Sustainability transition · Socio-technical map

5.1 Introduction

The aim of the Freiburg case study is to show that the ST-map may be useful to analyze the sustainability transition that has transformed an emerging 'Auto-city' into a modern 'Eco-city'. This transition pathway is characterized by different actors (such as local authorities, farmers, clergy, students, tourism operators, etc.) that support the vision of a sustainable city and foster the urban policies aimed at reducing the use of energy and private cars. Along the pathway the main transformative mechanism in place is based on the clustering of existing and emerging niches and systems of low-carbon mobility, such as public transport, individual bicycles and pedestrians. Sharing schemes are not particularly relevant in the Freiburg case.

G. Marletto et al., *Mapping Sustainability Transitions*,
SpringerBriefs in Business, DOI 10.1007/978-3-319-42274-9_5

Through this case study we can demonstrate that the sustainability transition towards the 'Eco-city' depends on the mutual interaction between the policies and practices of: urban mobility, land use, and energy. Moreover, we show that in this case citizens' participation and local authorities play a very important role.

The relevant literature and other information sources will be used to build a sequence of ST-maps, each representing a significant historical moment of the mobility system of Freiburg. In particular:

- mobility systems and niches will be positioned into the ST-map of Freiburg with reference to the cognitive elements that are crucial for any future innovation;
- for each ST mobility system and niche a supporting network of innovators will be identified;
- other external policies or events that influenced mobility system will be highlighted and analyzed, with a specific focus on land use and energy issues.

To help the readers to follow the sequence of relevant events a chronology is presented in the Appendix.

The sustainability transition process highlighted by the sequence of ST-maps will be discussed in the final paragraph of this chapter, also in order to identify any element that may be useful in achieving the sustainability transition of the mobility system in other cities.

5.2 The City of Freiburg

5.2.1 A Note on the Organization of the German State

Germany is a Federal Parliamentary Republic with four independent political levels: the Federal State, 16 Regions (*Länder*), 295 Districts and Cities not belonging to a District, and all other Towns and Municipalities belonging to a District. The local autonomy of Towns, Municipalities and Districts is a key element for the decentralized distribution of powers. These regional and local authorities have the right to govern themselves in relation to all issues concerning the local community matter. Towns and Municipalities exercise their legislative autonomy in the policy areas listed in Table 5.1, as long as they remain within the framework established by federal laws.

On the basis of the 1960 Federal Building Code of Germany that defines the basic conditions for urban planning all over Germany, cities have the autonomy to draw Land Use Plans and Zoning Plans. Moreover, cities can be the owner (or co-owner) of a power plant; only in this case cities may implement and independent energy policy. In the case of very big energy plants, representatives of the State, the Land and the City are member of the company's board. (Daseking 2016, PERSONAL COMMUNICATION)

Table 5.1 Competences of Towns, Cities not belonging to a district, and Municipalities

Administrative level	Competences
Towns, City not belonging to a District, And others Municipalities	• Urban planning • Municipal taxation • Public security and order • Municipal roads • Public transport • Road traffic • Registration of vehicles and vehicle taxation • Water supply and waste water management • Flood control and management • Fire fighting • Social aid and youth • Child care • Housing • School building and maintenance • Cemeteries • Other optional competences in the fields of energy, economic development, infrastructures, culture, sports, migration and integration

Source http://www.ccre.org/en/pays/view/7

5.2.2 *Basic Data*

The Baden-Württemberg Land is divided into four Districts: Freiburg, Karlsruhe, Stuttgart, and Tübingen. The Freiburg District is divided in nine Rural Districts and one independent City: Freiburg im Breisgau (now on, Freiburg).

Freiburg is a city of 153.06 km^2 of extension including a 40 % of forests. It is located in the southwest of Germany, at the edge of the Black Forest and a few kilometers from France and Switzerland. Freiburg has 225,000 inhabitants, including 30,000 students; its District has 650,000 inhabitants (Haag 2013).

Approximately 95,000 people work in Freiburg and 65 % of workers use the cars to move to or from the city (Beim and Haag 2016). The city has experienced over the years, a systematic growth of both outbound and inbound commuters (Table 5.2). The city also attracts about 3 million tourists per year.

Freiburg is quite flat and there are no significant barriers to cycling and to the operation of a tram network. Moreover, it is located along the Mannheim-Basilea (*Rheintalbahn*) railway line, one of the most used railway lines in Germany.

Table 5.2 Daily commuters to/from the city of Freiburg

Year	Inbound commuters	Outbound commuters
1970	34,000	3000
1987	53,000	8000
1997	64,000	14,000
2003	95,000	16,000

Sources Haag (2013), Beim and Haag (2016)

Table 5.3 Modal split of urban mobility in the city of Freiburg

Year	Walking (%)	Cycling (%)	Public transport (%)	Car (drivers) (%)	Car (passengers) (%)
1982	35	15	11	30	9
1989	24	21	18	29	7
1999	24	28	18	24	6
2001	24	28	18	30	

Source Banister (2005), Beim and Haag (2016), Freiburg City Council (1999)

Freiburg is also served by the A5 motorway Frankfurt-Basel and three federal roads (Beim and Haag 2016).

The evolution of Freiburg modal split shows how car use has been reduced in favor of public transport and cycling (Table 5.3).

As shown in Fig. 5.1, Freiburg is part of the small group of European cities with walking and cycling above the 50 % of the total mobility. Among these cities, only Breclav and Shkoder show a lower percentage in the use of private cars than Freiburg. However, it is important to consider that Breclav and Shkoder have a much smaller population than Freiburg.

Figure 5.2 shows that since 1990 Freiburg stopped following the same increasing pattern of car use of Germany and the USA.

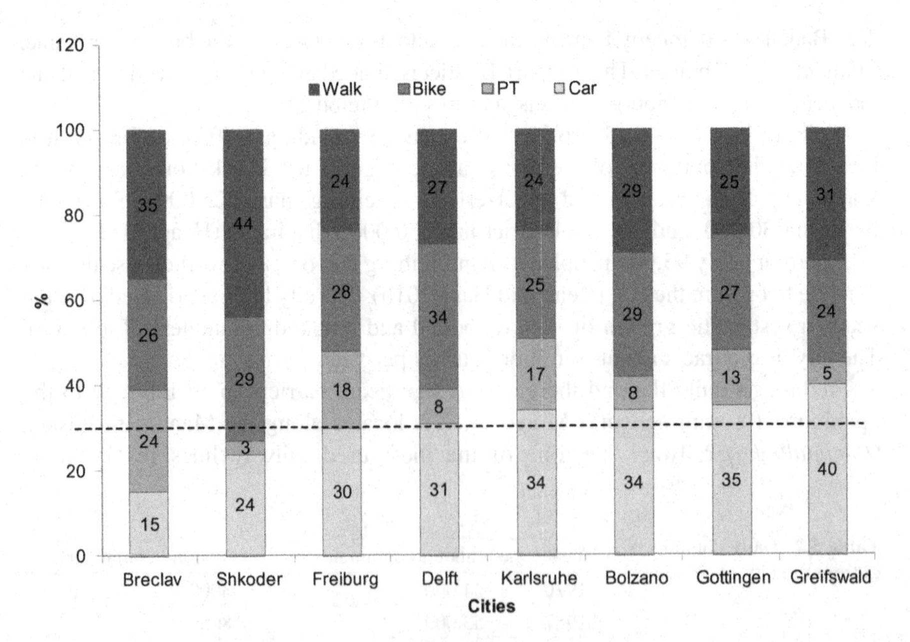

Fig. 5.1 Modal split of citIies where the share of cycling and walking is more than 23 %. *Source* TEMS, The EPOMM Modal Split Tool, http://www.epomm.eu/tems/result_cities.phtml?new=1

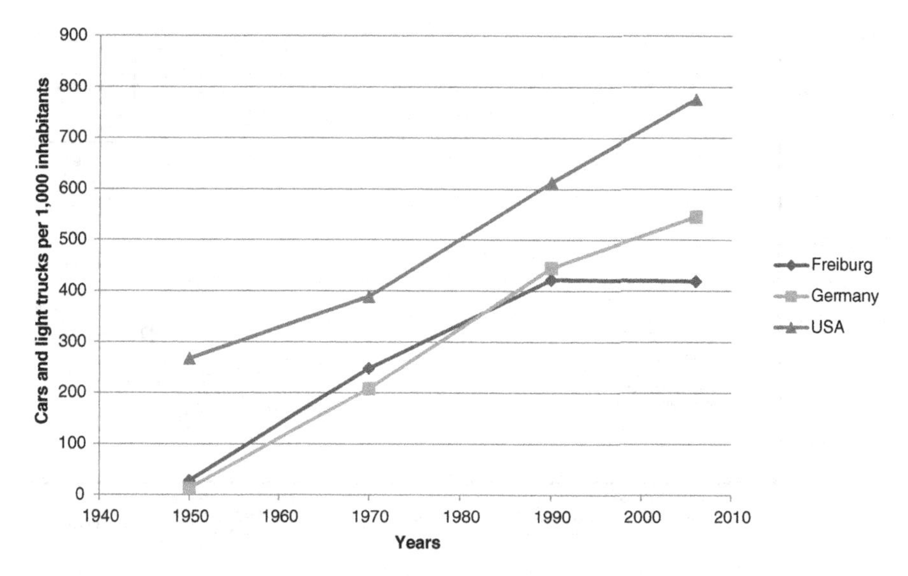

Fig. 5.2 Cars and light trucks in Freiburg, Germany and the US (per 1000 inhabitants). *Source* Buehler and Pucher (2011)

5.3 The Sustainability Transition of Urban Mobility

5.3.1 The Growth of Motorization (1945–1969)

We can find the basis for the transition pathway of Freiburg in a series of subsequent events beginning at the end of World War II. In 1944 the 80 % of Freiburg was destroyed by an air raid. The reconstruction began in 1947: all destroyed areas were rebuilt by following the principles of continuity with the past, and quality. Policies also promoted the use of private cars. Even if the traditional materials and design of buildings and medieval irregular narrow streets were maintained, the old historic squares were transformed in parking lots and a direct connection between the Autobahn and the city center was built. In 1949 in Freiburg there were only five small streets in the city center where the car could not enter; walkability was not considered as a constituent of the urban mobility system; pedestrians were not considered by policies and. (Lennard and Lennard 1995; Buehler and Pucher 2011; Medearis and Daseking 2012; Kelemen 2015)

Furthermore, after the end of the World War II, the *Wirtschaftswunder* (economic miracle) occurred: inflation was low and there was a rapid industrial growth. In this context, the population of Freiburg grew and it was necessary to build new settlements at the borders of the existing old city. As well as throughout Europe and the USA, the Freiburg 1955 Land Use Plan focused on the urban expansion made possible by the use of private cars. New settlements were characterized by large

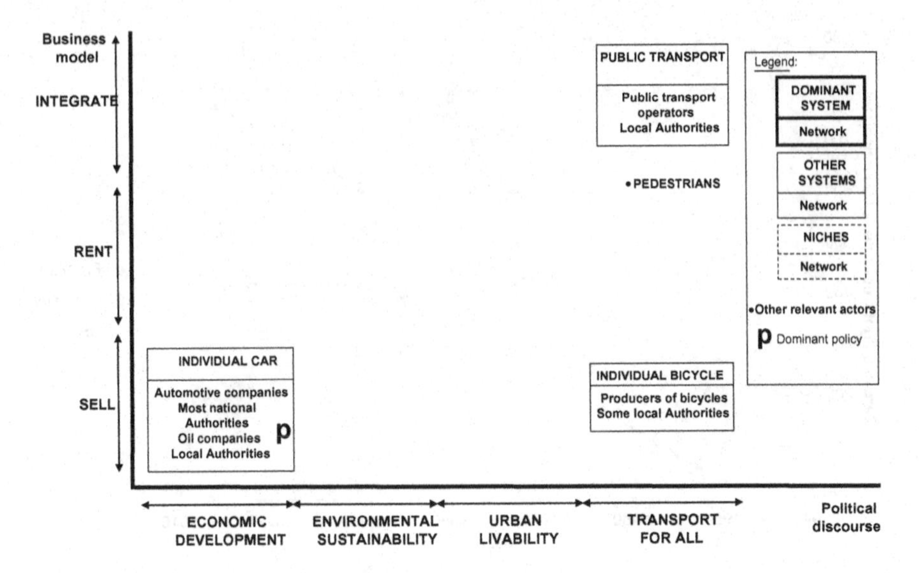

Fig. 5.3 The socio-technical map of the 1969 Freiburg mobility system

streets and parking lots. At the same time, tram lines are progressively abandoned and buses (though less efficient) were preferred. As a result, public transport was less and less important and—as the ST-map of 1969 shows—it no longer represented the dominant transport system (Fig. 5.3). At the same time, car ownership and use increased: in 1950, in Freiburg there were much more cars and light trucks than in West Germany (28 vs. 18 per 1000 inhabitants, respectively); from 1950 to 1970, air pollution, traffic fatalities, and traffic congestion increased. (Pucher and Clorer 1992; Greene 2004; Buehler and Pucher 2011; Kelemen 2015).

5.3.2 The Change Begins (1970–1979)

At the end of the 60s a change of direction affected Freiburg transport policies. The second Land Use Plan was never approved. It was car-oriented as the first: just for this reason a long debate among citizens, council members and administrations took place. In the 70s, it was finally shelved (Buehler and Pucher 2011). As a result, the city of Freiburg—unlike most other German cities—has never destroyed the historic center to improve the accessibility of cars: the modernist phase did not leave its strong footprint in the planning of urban structure (Beim and Haag 2016).

The first Integral Traffic Plan was drafted in 1969. In this plan, even if the dominant system of transportation was the individual car (Fig. 5.4), the needs of non-motorized inhabitants were respected. Another step into the direction of the 'Eco-city' model occurred in 1971: the Network of Cycling Routes Plan was drawn up and approved by the City Council. Since this first step cycling increased its

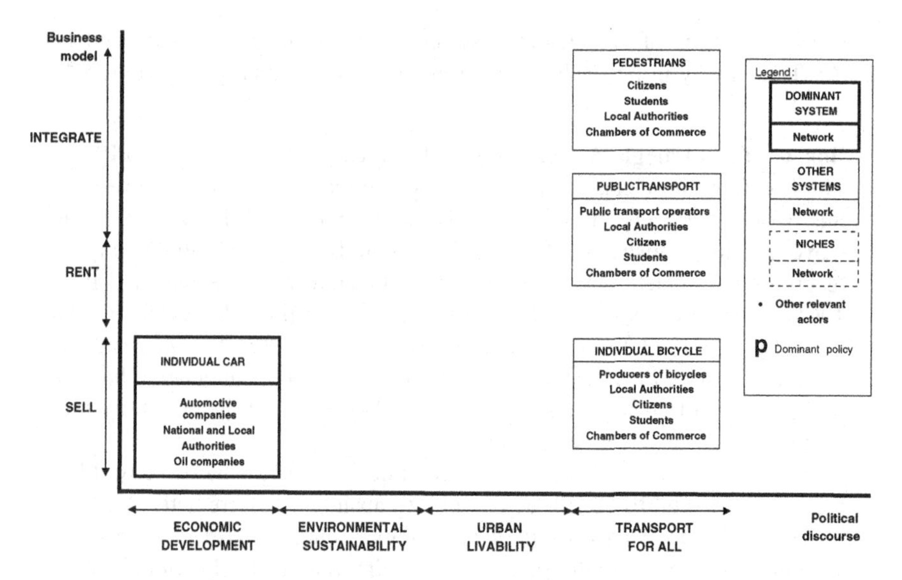

Fig. 5.4 The socio-technical map of the 1979 Freiburg mobility system

importance in Freiburg transport policy. In 1972, Freiburg municipality decided to maintain and expand the tram network; this action was based on modern concepts: separate tracks, priority traffic light junctions, a higher average speed. In 1973, the Freiburg city center has been transformed into a pedestrian area. This urban intervention helped to develop a new urban culture characterized by two elements: the preservation of the old town and the increase of cycling and walking. (Beim and Haag 2016).

The opening to pedestrian and cycling mobility, and public transport was due to several environmental and social problems caused directly by the car-based system. In addition, two events (the building of nuclear power plants and the 1973 oil crisis) brought public opinion to look with interest to the energy matter and to take side in a strong way in favor of a reduction of energy consumption in all sectors and a broad introduction of renewable energies. In this context, the policies based on land use change and car-oriented urban transport could not find a wide consensus. Furthermore, the participatory and inclusive model at the base of the struggle of Freiburg citizens fight against the Wyhl nuclear power plant (Box 1) had an overall influence on local political practices; in particular, the City adopted a policy approach based on collective discussion where citizens may take part in collective decisions. A very important role was played by the interaction between students and large organizations such as the Chamber of Commerce. (Patterson 1986; Bratzel 1999; Karapin 2007; Beim and Haag 2016).

In 1979 the Second Integral Traffic Plan was drafted and approved by the City Council. In this plan, unlike the previous one, pedestrians, bicycles and public transport had the same importance as individual cars, but still they were not considered as parts of an integrated system of urban mobility (Beim and Haag 2016). This is why in the 1979 ST-map there is no dominant policy (Fig. 5.4).

Box 1—The Struggle Against the Wyhl Nuclear Power Plant (NPP)

The first German nuclear plant was a research reactor that began to work in the town of Garching (near Munich, Bavaria) in 1957. In 1960 the German Government established the Germany's Atomic Energy Act with the purpose of promoting nuclear energy. Eight NPPs for civil use were activated from 1961 to 1969. In 1971 Wyhl was mentioned as a possible site for a new NPP. In July 1973 the Land government of Baden-Wurttemberg and the Southern Atomic Plant, a Land-owned utility company, announced the plan to build a NPP near the small town of Wyhl in the Rural District of Emmendingen in Baden-Wurttemberg.

The 1973 oil crisis pushed the growth of nuclear energy in Germany: 11 of the 35 German NPPs (8 still active and 27 abandoned) were activated from 1971 to 1979. In August of 1974 opponents, including farmers, clergy and students of nearby Freiburg, constituted the International Committee of Baden-Alsace Citizen Initiative. In the following years local opposition increased (Karapin 2007; Patterson 1986). The active participation of farmers clearly indicated that this was not a protest of the usual "leftist radicals" and but how it was rooted in the community. However, protests did not stop politicians and planners: official permission was granted and construction of the NPP began on February 1975 (Patterson 1986). Several hundred people reached and occupied the site, but were removed afterwards by a force of 700 policemen on February 20. This brought many supporters to join the emerging National antinuclear movement. On February 23, the site was re-occupied by 30,000 people. The occupation lasted until January 1976. In the meantime, in February 1975 the government stopped the construction of the NPP for several days. In March 1975, the Freiburg administrative court stopped the license and started the negotiation of so-called Offenburg agreement that was signed in January 1976. With this agreement the occupation ended. The agreement called to make detailed studies on the impact of the NPP on climate and the environment.

In March 1977 the Freiburg court suspended the construction of the NPP for 5 years. However, in 1982, the Government of Baden-Wurttemberg announced its intention to resume the project. As a consequence, the Baden-Alsace Initiative mobilized again with the same non-violent approach that was used in the 70s. The Land abandoned plans for the Wyhl NPP in 1983. In the 20th anniversary of the occupation of the site (1995), the regional government that the area where the NPP should have been built will instead be declared a natural reserve. (Patterson 1986; Karapin 2007).

5.3.3 The 'City of Short Distances' (1980–2010)

In 1989, the third Integral Traffic Plan was approved. It aimed at reducing car use through the promotion of environmental friendly transport modes and the implementation of some restrictions to car traffic (Beim and Haag 2016). This is the first Freiburg Plan that considers urban livability as the main policy goal. Only in the two subsequent Integral Traffic Plans the objectives became the urban livability and environmental sustainability (Fig. 5.5).

With the Integral Traffic Plan of 1999, Freiburg took a further step towards the 'Eco-city' (Freiburg City Council 1999). Besides reducing car traffic, it considered the overall transport system in which, public transportation, walking and cycling, worked in an integrated manner. Moreover, the Plan had the goal of contributing to environmental protection (Beim and Haag 2016).

Also in the Transportation Plan in 2002 the main goals were to improve urban liveability, environmental sustainability, public transport and accessibility conditions for pedestrians and cyclists. Transportation policies were integrated with urban planning and the above goals were calibrated to the conditions of population growth and regional development. The land use and transportation planning, over the last 40 years, have highlighted the importance of green mobility, public transport, traffic calming, mixed-use and human-scale urban development.

The reduction of car use in Freiburg reduction thanks to many different measures. The best known was the introduction, in 1984, of a low-cost monthly ticket called "urban environmental protection ticket" that turned into a regional ticket afterwards (FritzRoy and Smith 1998; Beim and Haag 2016). Around 86 % of all journeys by public transport in Freiburg are made by the owners of monthly or annual tickets

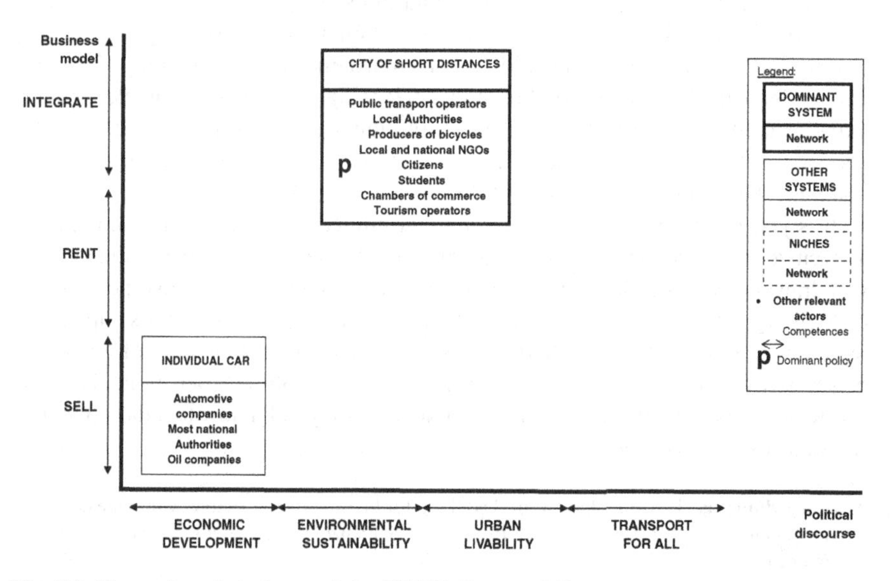

Fig. 5.5 The socio-technical map of the 2010 Freiburg mobility system

(Statistisches Jahrbuch 2009). Other measures (provided for in 2010) to increase sustainable transport included: traffic management policies, such as the further extension of the tram network, development of cycling infrastructure and improvement of walkability; land use policies, such as the better use of urban areas with brownfield investments and the functional mix of neighborhoods. (Beim and Haag 2016).

All the above plan and interventions have contributed to create the currently dominating integrated system of urban mobility also called the 'City of Short Distances' (Beim and Haag 2011) (Fig. 5.5).

5.4 Discussion and Conclusions

ST-maps of the Freiburg urban mobility system highlighted its sustainability transition from an emergent 'Auto-city' to a modern 'Eco-city'. Looking at the sequence of ST-maps, it is apparent an actual transition from a system characterized by an increasing role of individual cars (1969), to a system that aimed to a balanced role of different transportation modes (1979), and finally, to a system called the 'City of short distances' (2010), in which public transportation, walking and cycling are integrated.

Through the ST-maps a spatial divergence is also apparent between local actors that promoted the 'City of Short Distances', and national and international actors (e.g. the Federal Government, Oil and automotive companies, etc.) kept supporting the individual car. The divergent role of local actors is explained by two interacting dynamics. First: since 1960 the city of Freiburg has developed an integrated approach to several policy issues, such as urban growth, land use change, decarbonization, reduction of energy consumption, use of renewable energy and urban mobility. Such an integrated approach contributed to the gradual but continuous increase of citizens' awareness of the interaction between mobility and energy issues, and between urban livability and environmental sustainability (Hopkins 2009). Second: the struggle of Freiburg citizens against the Wyhl nuclear power plant generated a relevant change in institutional practices that resulted in the permanent integration of an increasing number of actors into political decision and planning (Patterson 1986; Bratzel 1999; Karapin 2007).

It is just under these conditions of mutual influence between policy integration and citizens' participation that a critical mass for change has been created: the withdrawal of the second—still car-oriented—Land Use Plan was just the first step of Freiburg from an 'Auto-city' to an 'Eco-city'. The number of actors supporting the today dominant 'City of Short Distances' signals that the creation of a network of supporting innovative actors is key for any sustainability transition; the composition of this specific network stresses that technological innovators did not play a relevant role in this case.

Acknowledgments I wish to thank Prof. Daseking for his stimulating lecture within the course of Urban planning held at University of Rome "La Sapienza", and for the following personal communications.

Appendix

See Table 5.4.

Table 5.4 Chronology of relevant plans and events in Freiburg: transportation, land use, nuclear energy (1947–1984)

Year	Plan	Event	Actors involved
1947	First land use plan		Freiburg Municipality
1969	First integral traffic plan is drafted		Freiburg Municipality
1970	Second land use plan is shelved		Freiburg Municipality
1971	Network of cycling routes plan is drafted		Freiburg Municipality
1972		Decision to maintain and expand the tram network	Freiburg Municipality
1973		The city center is transformed into a pedestrian area	Freiburg Municipality
1973		Project of a new nuclear power plant (NPP) in Wyhl	Land Government Southern Atomic Plant
1975		Construction of the Wyhl NPP begins	Land government Southern Atomic Plant
February 1975–January 1976		The site of Wyhl NPP is occupied	Farmers Students of Freiburg University Other citizens adverse to the Wyhl NPP
1977		The construction of the Wyhl NPP is suspended	Freiburg Court
1979	Second integral traffic plan		Municipality of Freiburg Students Citizens
1982		The project of the Wyhl NPP is resumed	Superior Court Baden-Wurttemberg Government
1982		Non-violent protests against the construction of Wyhl NPP	International Committee of Baden-Alsace Citizen Initiatives
1983		The project of the Wyhl NPP is abandoned	Baden-Wurttemberg Government
1984		Environmental ticket	Freiburg Municipality

References

Banister D (2005) Unsustainable transport: city transport in the new century. Routledge, London and New York

Bratzel S (1999) Conditions of success in sustainable urban transport policy—Policy change in 'relatively successful' European cities. Transp Rev 19:177–190

Beim M, Haag M (2011) Public transport as a key factor of urban sustainability. A case study of Freiburg.https://www.researchgate.net/publication/264651721. Accessed 20 April 2016

Beim M, Haag M (2016) Freiburg's way to sustainability: the role of integrated urban and transport planning. http://www.corp.at/archive/CORP2010_56.pdf. Accessed 04 April 2016

Buehler R, Pucher J (2011) Sustainable transport in freiburg: lessons from germany's environmental capital. Int J Sustain Transp 5:43–70

Lennard SHC, Lennard HL (1995) Livable cities observed: a source book of images and ideas. Gondolier Press Book, New York

Daseking W (2016) Green neighborhood in Freiburg. Lecture held within the course of Urban Planning. University of Rome "La Sapienza"

Freiburg City Council (1999) Verkehrsentwicklungsplan (Transport development plan). http://www.freiburg.de/pb/site/Freiburg/get/documents_E-794160118/freiburg/daten/verkehr/vep/VEP%20Analysebericht.pdf. Accessed 30 March 2016

FitzRoy F, Smith I (1998) Public transport demand in Freiburg: why did patronage double in a decade? Transp Policy 5:163–173

Greene G (2004) The End of suburbia: oil depletion and the collapse of the American dream. The Electric Wallpaper Company, Canada

Haag M (2013) Sustainable Transport Initiatives in Freiburg. EcoMobility Congress, Suwon. http://ecomobility2013.iclei.org/fileadmin/user_upload/ecomobility2013/Presentations/P2_Haag.pdf. Accessed 26 May 2016

Hopkins R (2009) The Transition Handbook: from Oil Dependency to Resilience. Totnes, Green book

Karapin R (2007) Protest politics in Germany: movements on the left and right since the 1960s. Penn State University Press

Kelemen RD (ed) (2015) Lessons from Europe? What Americans can learn from European public policies. CQ Press, Thousand Oaks (CA)

Medearis D, Daseking W (2012) Freiburg, Germany: Germany's Eco-Capital. In: Beatley T (ed) Green Cities of Europe-Global lessons on green urbanism. Island Press, Washington

Patterson WC (1986) Nuclear power, 2nd edn. Penguin Books, Harmondsworth

Pucher J, Clorer S (1992) Taming the automobile in Germany. Transp Q 46:383–395

Statistisches Jahrbuch (2009) Beiträge zur Statistik-Amt für Bürgerservice und Informationsverarbeitung. Stadtamt Freiburg, 2009

Chapter 6
The History of the US Light Market Between Bulbs and Illumination

Abstract This chapter combines a historical case study with the ST-map which aims at describing the evolution of the main U.S. light systems over the last two centuries. The use of the ST-map helps to track the evolution of new technologies—and their networks of innovators, and it helps to position such networks in respect to the relevant business models and the dominant political discourses. The analysis represents how the dominancy of specific light systems has been intertwined with the development of networks of innovators that could seize value from the value of new urban energy infrastructures first, and later from the growing importance of the energy/environmental question in the political discourse about light. The ST-map represents a simplified way to present the dynamics of dominant systems over long time, so it may be used to convey important dynamics—in a synthetic way—overcoming a limitation of typical historical descriptive case studies. This chapter also addresses the limitations encountered by the usage of ST-map in historical case studies, especially when the focus is on narrow socio-technical systems.

Keywords Lighting · Socio-technical system · Business model · Political discourse · Network of innovators · Socio-technical map

The following case study presents the evolution of the U.S. indoor light market during the period 1880–2030. The analysis is based on a combination of existing sources (e.g., reports, papers) for the historical part and interviews with experts for the future-looking part.[1] Five main periods are identified and an ST-map is presented for each period. The proposed ST-maps represent the most relevant networks of innovators and their dynamics along two dimensions: the political discourse and the business model. The case ends with some considerations about the usage of the ST-map as analytical tool.

---

[1]For detailed references, see Franceschini and Pansera (2015) and Franceschini and Alkemade (2016).

G. Marletto et al., *Mapping Sustainability Transitions*,
SpringerBriefs in Business, DOI 10.1007/978-3-319-42274-9_6

6.1 The Pre-electrification Era (up to 1870s)

The development of the first illumination systems was deeply linked to the support of policy makers which saw light as a way to increase safety and security of urban areas. First examples of such policies date back in Europe at the laws of London (1417) and Paris (1524) where it was ordered that citizens must hang lamps with light at windows during the night so the people walking outside could see at night.

In the 1870s, the first electric lamps (Edison and the arc-lamp) became available, but, up to that period, the provision of light was mainly fulfilled by the gas lamp and the oil lamp. The gas lamp was considered superior in respect to efficiency and cleanliness—essential elements for indoor application—but it was suffering for some safety issues connected with the risk of suffocation or explosions, and it was a fixed lamp. The oil lamp was portable and did not have the risk of suffocation or explosion, though it was not long lasting and the burning process was pretty polluting.

Figure 6.1 positions the networks of innovators for both gas and oil systems, representing the main actors and the relevant business models for both systems.

The oil lamp had a dramatic improvement thanks to the invention of kerosene. Before it, whale oil and tallow were the most relevant resources for oil lamp but the increasing demand made such resources very expensive. The invention of kerosene heavily reduced the running cost of the lamp making it a cheap source of light.

As showed in Fig. 6.1, the oil light business model was essentially based on the sale of a lamp, while the sale of the gas lamp was actually part of an overall development of the gas infrastructure. Therefore, gas players were focusing on the

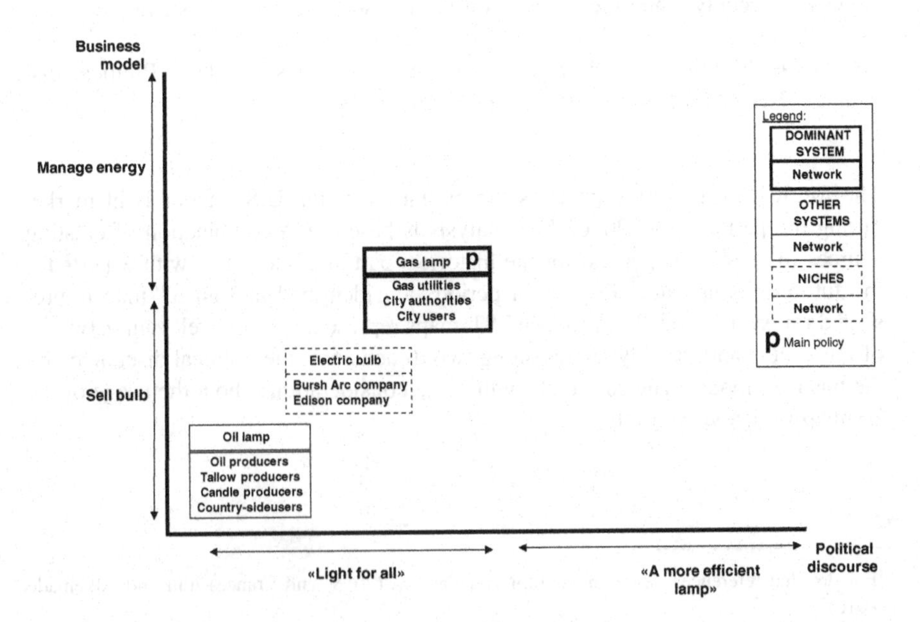

Fig. 6.1 Socio-technical map of the US lighting market (1880)

sale of gas, whose gas lamp was part of, through the development of new gas infrastructures which could have been used also to heat and cook—getting the favor of policy makers and citizens. Gas pipes were laid in cities and gas lamps were positioned on the posts. Baltimore is recognized as the first U.S. city to have installed a gas light system in 1816. When the gas infrastructure was mature—only in the second half of the century—the gas lamp could prevail and become dominant. Oil lamp—based on kerosene—kept important market positions only in rural areas, where gas infrastructures were absent and oil wells—used for the production of kerosene—were developed.

6.2 The Electrification Era (1880s–1930s)

The arc-lamp and the Edison lamp—the first two electric lamps—were both commercialized in the late 1870s in the context of growing expansion of the gas lamp. The arc lamp—the first one to be available on the market—was very brilliant, but it suffered of many limitations. Even the smallest arc lamp was too powerful and brilliant to be used in many market segments, such as the indoor one. In addition, the arc lamp had a relevant energy consumption and would have required an adequate electric system that was not yet developed. For these reasons, the arc lamp found applications only in special industrial settings such as in lighting houses, and only for some outdoor applications.

The Edison lamp played a pivotal role in diffusing the electric lamp, because it targeted indoor applications—a segment which could not be fulfilled by the arc lamp—that were in need of light. While the gas lamp was still the most convenient, and the arc lamp the most effective, the Edison lamp provided a simple and adequate solution for indoor light even though it was the least economical of the three. For two more decades, the three technologies co-existed: the gas lamp and the arc lamp were mainly competing for the outdoor market; the incandescent lamp and the gas one were competing for the indoor market. The competition between the different technologies was represented by the different discourses about light. When the incandescent lamp was available, its supporters focused on the quality, versatility and customizability of illumination that made such technology ready to be used in any context. Against them, the gas supporters pointed out the cost of the electric lamp and the impossibility to provide widespread illumination given the required amount of energy.

The competition in the lighting market deeply changed with the development of the tungsten incandescent bulb in the 1910s which replaced the first carbon incandescent bulb of Edison. The tungsten incandescent bulb outperformed all the existing competitors in respect of efficiency and luminosity. The combination of the electrification of the market with the tungsten technology led to a radical simplification of the light market, in which the tungsten incandescent bulb became the dominant solution for both indoor and outdoor applications. Gas light players realized the superiority of the tungsten bulb. Some of them entered the electric light market, but the majority turned their attention to the provision of gas for heating, cooking, and cooling, leaving the light market.

The market showed a strong process of consolidation as well, with the birth and dominance of General Electric (GE) from the fusion of Edison General Electric Company and Thomson-Houston Electric Company.

Figure 6.2 shows the new configurations of networks of innovators by the end of 1930s. Four elements deserve attention. First, the gas and oil light systems disappeared as the result of the new tungsten bulb. Second—and similarly to Fig. 6.1—the dominant system had an important component of the business model based on the management of energy—electricity in this case. Third—and again similarly to Fig. 6.1—the new dominant light system has been backed by favorable policy makers. Fourth, a new competitor—that is the fluorescent light—appeared in the market. While the first element has been already discussed, the rest of this section focuses on the other three elements.

By the end of 1880s, electric lamp players understood the need to provide reliable and cost effective energy system in order to back the diffusion of the electric light. In fact, the diffusion of electric lamp was hindered by the lack of an adequate system of generation and distribution of electricity. Electric players were aware that the development of the gas lamp was connected to the development of a new gas infrastructure and that the new electric lamp had the same characteristic. Therefore, many electric lamp players moved in the production and diffusion of electricity. The result was that both the incandescent and the arc lamps were deeply connected to the electrification of urban areas. While the telephone and telegraph were actually

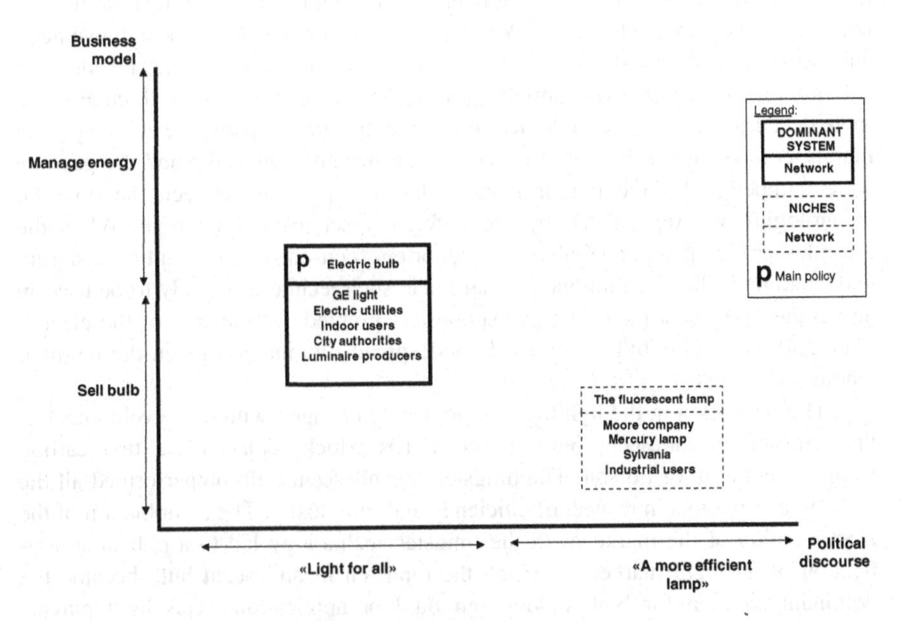

Fig. 6.2 Socio-technical map of the US lighting market (1940)

the first two technologies which brought electricity in urban areas, these technologies had so limited energy intensity in respect to the need of electric light that did not encourage the development of a strong and cheap electric infrastructure which would have supported the widespread diffusion of the new light.

The development of the electric lamp was in fact the development of a service-based business model based on the sale of an electric lamp *and* the provision of electricity, initially based on local and isolated energy systems. It was such combination which made the electric lamp competitive in respect of the gas one in any market segments. Among many examples, Edison launched "The Edison Company for Isolated Lighting" in 1882 with the purpose of promoting small power plants to be settled in stores, offices, hotels and other non-residential settings. Already in 1886, there were 702 small isolated generation plants which provided light to 181,463 lamps (Bright 1949). Similarly, the Bursh arc-light system installed the first electric central station for urban areas. Even though the arc lamp and the incandescent lamp could not initially be used on the same electrical system, both these technologies were deeply connected to the electrification of the U.S. market.

Municipalities and local authorities played a pivotal role in the diffusion of the electric light, because they strongly supported the development and diffusion of a stable electric grid. The result was that both isolated and centralized energy systems had a strong institutional support, making the incandescent light the most prominent light solutions for urban areas. The city of Wabasha, Indiana, is actually recognized as the first electrically lighted city in USA. In 1880, the city council decided to have a first illumination system based on the arc lamp.

During this period, a new competitor—the fluorescent lamp—was also under development. The first competitor—the Moore tube—was developed in competition of the Edison Lamp that was considered "*too small, too hot and too red*" (Moore as cited in Bright 1949, p. 221). The first Moore tube had an efficiency nearly double than the incandescent bulb, but it was very expensive, complicated to install and high voltage demanding. The Moore tube had a discrete success before the tungsten bulb appeared in the market. Since that time, the Moore tube lost its advantage, disappeared from the market, and that knowledge was absorbed by GE which was interested in the fluorescent light.

6.3 The Co-existence Period (1940s–1960s)

The 1940s was a turning decade for the light market, because of the commercialization of a new fluorescent tube light. In fact, the dramatic popularity of the incandescent bulb created some relevant issues for the light system (Inman 1939): (i) dramatic increase of the energy bill for non-residential users; (ii) electrical wiring overload with several black-outs; and (iii) excessive heat dissipation which made uncomfortable many non-residential indoor environments.

Experts and professional users knew the research and progresses of the fluorescent technology and the potential benefits of developing more energy-efficient light sources. They demanded to the most relevant lighting players to present new prototypes at New York's World Fair in 1938. The presentation was an unexpected success. The fluorescent tube was nearly three times more efficient then the incandescent bulb, and this performance attracted the interest of the non-residential markets, such as offices, retailers and commercial centers, where the incandescent bulb was massively used.

GE Electric was worried that widespread diffusion of fluorescent light could damage the interest of its network formed by electric utilities—which had their business based on the sale of electricity—and of the luminaire producers—whose products were not compatible with the new tube. Therefore, GE' aim was to focus on the commercialization of the incandescent bulb, and to develop two defensive strategies for the fluorescent technology: (i) presenting the fluorescent tube to consumers as a special colorful lamps with no general purpose (Bijker 1995); and (ii) developing and patenting new knowledge about fluorescence. Actually some competitors, among which Sylvania was the most important, were able to develop their own fluorescent tube and started an aggressive marketing strategy to try to weaken the GE's market power. Sylvania quickly obtained 20 percent of the new fluorescent market, thanks to the widespread diffusion of the tubes among non-residential users, becoming GE's first relevant competitor.

The history of the 1940s–1960s is actually a period of co-existence of these two light systems which fulfilled very different needs of the residential and non-residential indoor market segments.

Figure 6.3 represents the light market in 1970. Two systems dominate—the tungsten and the fluorescent ones—and a new one is about to emerge: the LED system. The incandescent lamp focused on quality of light and fixtures, and versatility and cheapness of the installation. Residential users—which were more concerned about quality and beauty of light and luminaries than energy cost and intensity—preferred the incandescent bulb. The non-residential users—which were more disturbed by the energy efficiency limitation of the incandescent bulbs—adopted the fluorescent tube because of its better energy efficiency which: (i) improved thermal comfort—especially during daylight hours, (ii) reduced energy bill, (iii) increased reliability of electric systems. Sylvania understood that the first tubes lacked of appropriate fixtures and promoted the new "Miralumes" fixture. In fact, Sylvania sold not simply a tube, but a new complete and energy efficient illumination system for the non-residential market which included both tubes and fixtures. Interesting, this period shows that the birth of the efficiency issue in connection with light was not the result of a special political commitment, but started from the need of specific non-residential users which were concern with the increased consumption of electricity.

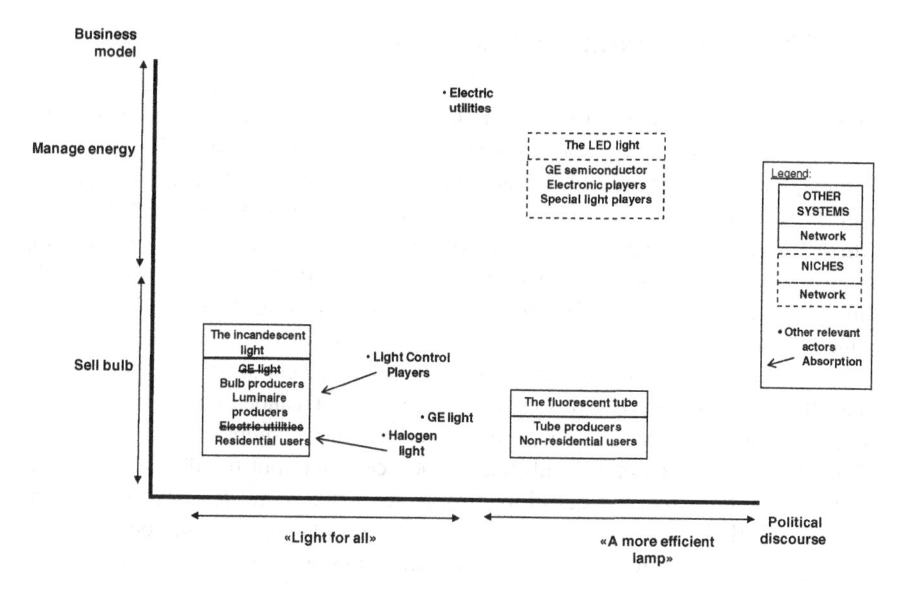

Fig. 6.3 Socio-technical map of the US lighting market (1970)

This long period was characterized by the dominance of a business model based on the sale of lamps. The sale of the electric lamp was not anymore so deeply connected to the provision of generation and distribution of electricity. Light players and electric utilities evolved as different players with their own business models. The different evolutions of the business models between light players and electric utilities made also possible for GE to slowly move in the fluorescent system. Once electric utilities' main business models were not anymore so linked to the dynamics of the light market, GE had more freedom and moved in between the two main light systems in order to have an influence in both the technologies: the incandescent one—which was high profitable—and the fluorescent tube—considered the future of light.

Finally, the 1960s is the decade of the birth of the LED light. LED was not yet framed as a light technology and its development was driven by the electronics industry seeking for an energy efficient technology to develop new electronic appliances. The major breakthrough occurred in 1962, when GE's rectifier department announced the invention of both the first infrared and the first visible red LEDs. Interesting, at that time, the LED was not yet considered a viable light source. In fact, GE's rectifier department did not have any connection with GE's light department in this first stage of development of the LED technology. Instead, by 1962, the electronics industry started very important R&D investments to develop this new technology, and this technology appeared in some markets, but not in the light one.

6.4 The Energy Issue (1970s–2000s)

The pacific co-existence of the two markets started being under pressure since the 1970, when the energy question became central in the U.S. political agenda. The first major power blackout in 1965, and the oil crisis in the 1970s, which culminated with the declaration of national energy supply shortage in 1979, were all factors which questioned the stability and security of the U.S. energy system. Even if the provision of light was not an explicit target of policies in that period, increasing the efficiency of the energy system became the most appealing solution for the U.S. policy makers. Many lighting players started to foreseen the value of promoting new energy efficient lighting technologies, and the residential market was the most appealing segment because of the dominance of the incandescent lamp. The light players started working on the development of a new compact fluorescent lamp (CFL) which could compete with the incandescent one, but the dominance of the incandescent bulb was not questioned for at least two more decades, because the fluorescent tube was still inferior in respect of light quality, size and beauty and it could not easily be fitted to be used with luminaries with screw-based sockets.

The 2000s represent a relevant technological shift for the residential market, because of the co-occurrence of specific energy and environmental policies focused on promoting energy efficient lighting and of the maturity of the compact fluorescent lamp technology. California was the front runner among the U.S. states. CFL market share quickly rose from 1 percent in the final quarter of 2000 to 8 % at the beginning of 2001 (Calwell et al. 2001; Iwafune 2000). The increasing share of the lamp was the result of the combination of new light policies, new interoperability, new standards, and evolution of CFL light quality. The California State was in the middle of a severe energy crisis, with 173 days of energy emergency only in 2001. The California state launched several initiatives to boost the CFL residential market in the 1990s and 2000s. Such policies promoted more attractive CFL lamps, competition in the industry, awareness among consumers, a strong public demand, and a ban for some incandescent applications. Only one decade afterwards, there were 100 CFL manufacturing partners with the *Energy Star* label on the overall U. S. market, producing 1600 unique CFL lamps (D&R International 2010). Latest market analyses indicate that almost 20 % of residential installed lamps in 2010 are CFLs (Swope 2010; Navigant Consulting 2012).

Figure 6.4 shows the positioning of the networks of innovators in 2010, when the fluorescent light system became relevant in the residential market too. This period is characterized by the dominance of a business model based on the sale of lamps, as it was in the previous period. It is worth to notice a marginal but increasing importance of a new class of building and energy players and technologies for light, which make the map more complex. Such new players—as energy management companies and energy saving companies (ESCO)—have been

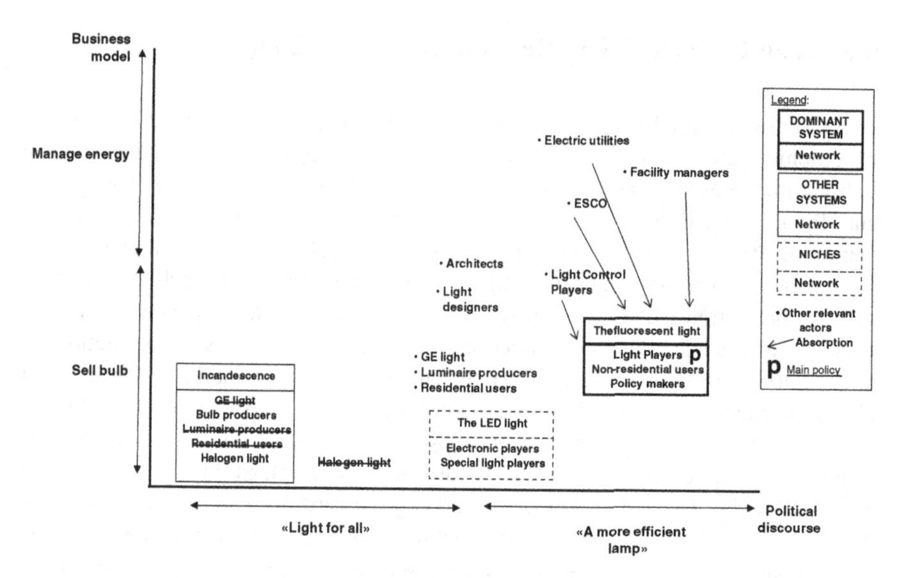

Fig. 6.4 Socio-technical map of the US lighting market (2010)

in the market since the 1970s, but their importance has grown only since the 1990s. These companies have promoted a management-based business model which is different from the dominant one for the last five decades.

As Fig. 6.4 shows, some relevant players—among which GE electric—were in the middle of a transition between the two systems with the result of weakening the incandescent light system in favor of the fluorescent one, which was backed by the dominant political discourse. In order to understand the reason of such shift, it is worth to remember that energy efficiency and energy security became important in the U.S. political discourse since the 1970s, as the environmental agenda did since the mid-1980s. The transition towards more efficient (light) technologies had a very different impact on the two main market segments. In the non-residential one—where energy efficiency was already in the political agenda—the main dynamics and technologies did not change, but there was only an increasing presence of energy related players. In the residential market, the new energy effort resulted in the incandescent technology being replaced by the fluorescent technology.

These three decades show also the first attempts of the LED technology to acquire specialized shares of the non-general light market. In the 1970s, the first yellow LED was announced. Since the 1980s, first LED applications started being sold in markets where the white color was not needed, such as disco-lighting systems, automobile central brake lamps, and traffic lights. Finally, in the mid-1990s, the blue LED was announced, opening the possibility to develop the white LED, though this technology was considered inferior by the light industry in respect to the potentiality of the fluorescent light.

6.5 The Current LED Revolution (2010s–2020s)

The LED light started appearing as a relevant competitor only in the 2010s, but its development process had a breaking point in 1999, when HP and Sandia National Laboratories presented revolutionary predictions about the LED technology (Haitz and Tsao 2011). Before 1999, LED technology was expected to achieve an efficiency of 50 lumens/watts by 2010 which made such technology interesting for some light applications, but not competitive with the fluorescent technology. With the new predictions, LED was expected to achieve up to 200 lumens/watts in only two decades and dramatically reduce its manufacturing costs. Such predictions—lately known as "Haitz's law"—deeply changed the light market. All the light players wanted to be part of the new technology that was expected to overcome the fluorescent technology within two decades. Since 2000, all the main light players acquired relevant electronic and semiconductor capabilities, essential to develop the new LED light. The new LED light market gave birth to a market in which traditional light players faced new competitors coming from the electronics and semiconductor industries, making such market highly competitive. The new LED promise attracted also the attention of policy makers which turned their attention from the fluorescent technology to the LED one. Private and public R&D investments nearly doubled compared with forecast investments in pre-1990s industry reports (Haitz and Tsao 2011).

At the same time, the ESCO market shows important signs of grow, with annual sales that increase about 9 % each year since 2009 (Stuart et al. 2013). The market revenue reached about $7.5 billion in 2014, and it is expected to get up to $10.6–15.3 billion in 2020. ESCO's strategies to achieve energy saving include the provision of more energy efficient light systems. Such strategy both includes more energy efficient lamps and better lighting management systems. The conceptualization of light as system has driven the diffusion of automatic controlling systems (ACS) for light. ACS have been commercially available for many decades, but such technologies have been marginal for long time (CEE 2014). Today only 2 % of residential lamps and 4 % of commercial lamps have a light or motion sensor installed (Navigant Consulting 2012), though can achieve saving of energy up to 80 %.

The increasing importance of LED technology and energy management is a precursor of a potential pathway for the future light system. Current industry expectations indicate that the future light system will be *smart* and heavily dependent on the LED technology. LED technology will achieve up to 200 lumens/watts of efficiency, but this technology will also exploit many other characteristics (e.g., better versatility, dimmability, quality of light) that are superior in respect to the current fluorescent light solutions. Such characteristics will enable a full integration of light control systems which can take advantage of the versatility of LED. Domotics is expected to quickly develop as a new market in which light management will be an essential component. As a result, the future *smart* LED light system will open new types of applications that will be fully exploited only when new competences—which come from light designers and architects—will be integrated in the light market.

A further area of development is the organic LED (OLED) which is expected to bring a fully new light experience. The OLED technology is expected to make virtually any surface a light emitter. Future OLED light system may not be based on a lamp as emitting source of light—as it happens today—but they take the form of light coming from any surfaces over which a thin layer of OLED is applied. The OLED technology is therefore expected to encourage the transition towards illumination as a system and not being anymore lamp-centered.

The future light system seems quite radical also in terms of changes to dominant light discourse and business model (Fig. 6.5). Today energy efficient/consummation is the central narrative. Developing more efficient light technologies has been the core activity of light players in the last decades. In the future, other dimensions such as comfort, aesthetic, health and productivity, are expected to become more important in the assessment of light performances. This change is the result of a combination of increasing knowledge and awareness about the effects of light on health and human productivity, and of the increasing versatility of LED-based light systems. Such dynamics may promote a more complex lighting discourse that will divert the future attention of lighting actors from energy saving to a more complex set of objectives. Saving energy may turn to be not anymore the priority for the sector, as soon as new objectives will arise in the light agenda. It is yet unclear whether the policy makers will support such new objectives or they will still focus only on energy efficiency. While there has been a strong agreement between policy makers and industry players in the last four decades, the new LED opportunities may put policy makers and industry players on two different sides, in which the first may still focus on energy efficiency, while the second promoting a more complex combination of objectives for future light systems.

New business models may arise from these shifts about technologies and discourses. Selling lamp is expected not to be anymore the core of the light sector. Future business models will seize the new LED value, by selling and managing light systems which include a complex set of controls, technologies and procedures. Since the illumination performances are heavily dependent on the environment in which they apply, the future light system is expected to promote a more functional service-based approach of light management that will fully integrate energy management and building management competences. The management of light will emerge as a new business model which integrates energy aspects with the different sources—natural and artificial—of light.

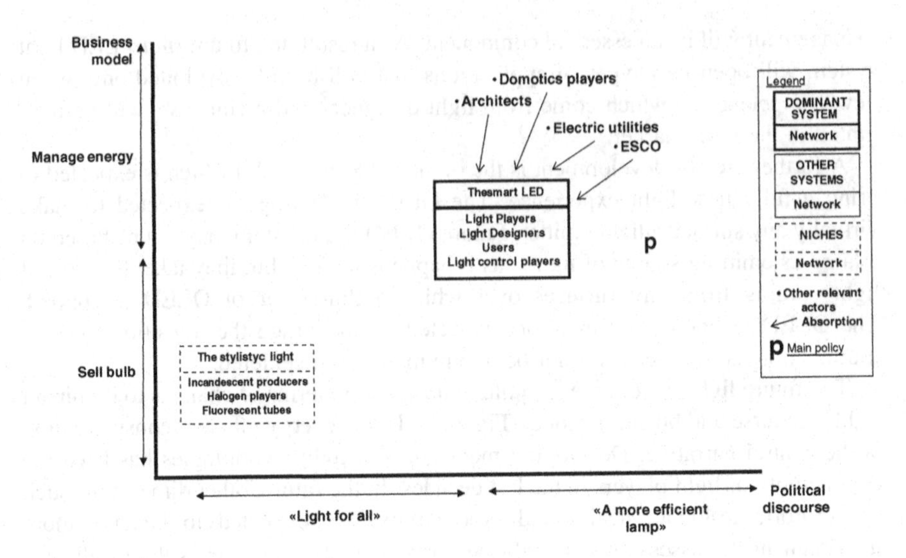

Fig. 6.5 Socio-technical map of the US lighting market (2030 scenario)

6.6 Final Remarks

In this chapter, I have used the ST-map to analyze the dynamics of the light sector in USA, a topic which I have already investigated (Franceschini and Alkemade 2016; Franceschini and Pansera 2015). This experience aimed to test whether the use of the ST-map can provide novel information—especially about the dynamics of the networks of innovators—in historical case studies. The following main considerations arise.

First, the ST-map has helped to highlight that different business models have been used over the time—shifting from a management-oriented approach to a product-based approach and back-, and that the success of the different systems was partially linked to the promotion of such business models. This aspect may have been underestimated in traditional analyses which compared lighting technologies *vis-à-vis*. The ST-map forces researchers to position the players in a two-dimensional map, so it stresses the need to understand differences and commonalities of the different networks of innovators according to the political discourse and techno-economic capabilities- here represented by the business models. Figure 6.6 compares the dynamics of the dominant ST-systems through the different periods.

The light sector started from a management-based business model approach, essentially linked to the development of the gas (up to 1870s) and the electric (1880s–1930s) infrastructures. Afterwards, the ST-map shows that the dominant business model turned towards the sale of bulbs (1940s–1960s and 1970s–2000s). Finally, in the 2030 scenario, it seems that the dominant business model will shift again towards a management perspective, based on the provision and management

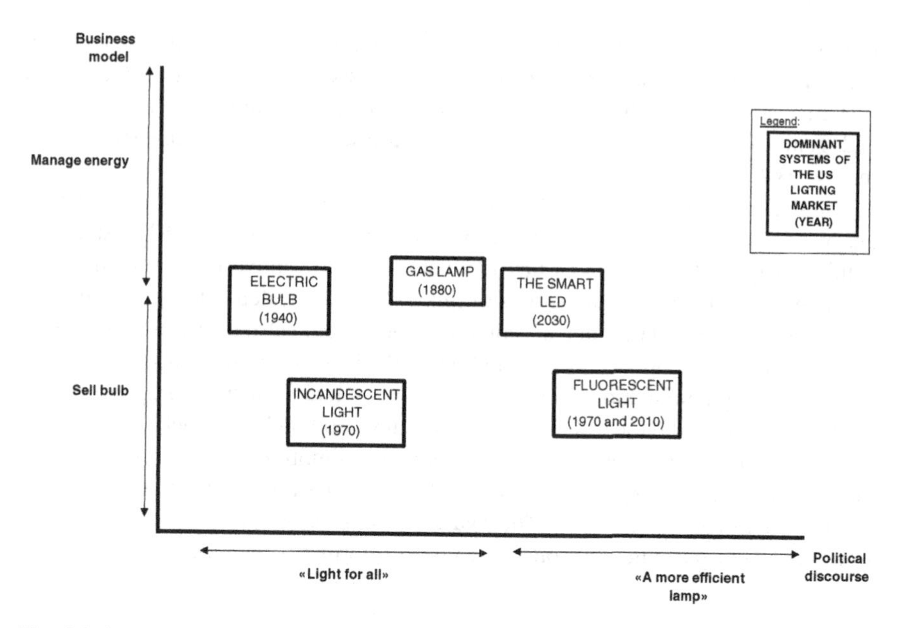

Fig. 6.6 Socio-technical maps of the US lighting market: a synthesis (1880–2030)

of illumination (2010s–2020s). A similar loop occurs for the political discourse as well. The dominant political discourse started with the provision of light (up to 1870s and 1880s–1930s), and it had a shift towards the concept of efficiency (1940s–1960s and 1970s–2000s). Finally, it seems that the main light discourse may move back in the future (2030 scenario) to the provision of light to people, as represented by the frame of the future smart LED light as a totally new sensorial experience.

Second, the use of the ST-map has highlighted a correspondence between the existence of a dominant ST-system and of a dominant favorable political discourse. The period 1940s–1960s is the only case in which there was neither a dominant light system nor a dominant political discourse about light. Therefore, it seems that the light case is an example in which the dynamics of the innovators and the dominant strategies are deeply related to the existence of a favorable political context. The descriptive analysis seems to explain that the dominant political discourse is the *cause*, while the dominant light system is the *effect* in this relationship. In fact, the dominance of the gas lamp in the period up to the 1870s and the overtaken by the electric lamp in the period 1880s–1930s is linked to the favorable policies to develop urbanization which encouraged the diffusion of the gas and electric infrastructures, respectively. Similarly, the arise of the energy and environmental questions in the period 2010s–2020s broke the existing equilibrium between incandescent and fluorescent lamps in the period 1970s–2000s. While in other contexts it may be expected a different causal relationship or even the impossibility to define that, an explanation for this case may be that the light system is too narrow to understand the evolution of complex political discourses. This explanation may suggest that widening the analysis (e.g., considering the energy system instead of the light one) may shed light in how political discourses are endogenously influenced. In other words, the case seems to suggest that the ST-map may work better with wider ST-systems.

Similar conclusions may be drawn by the difficulty to follow actors through the periods. It was difficult to represent the dynamism of innovators when ST-systems are continuously rearranging and when innovators seem to move in and out from the ST-map through the different periods of time. This may be dependent on old players exiting the market and new players coming in, but it could also be the case of relevant players which developed strategies and networks beyond the light system and on a wider level (e.g., production of energy). This case—again—seems to suggest that the ST-map may work better with a wider (and future-looking?) approach.

Finally, the ST-map provided a convenient and simplified way to show the evolution of the dominant systems over a very long period of time. This is an interesting quality which can increase the added value of long historical qualitative analyses, since the ST-map can help to convey synthetic information in a usable way.

References

Bijker WE (1995) Of bicycles, bakelites, and bulbs: toward a theory of sociotechnical change. Massachusetts Institute of Technology, London

Bright AA (1949) The electric-lamp industry: technological change and economic development from 1800 to 1947. The MacMillan Company, New York

Calwell C, Zugel J, Banwell P, Reed W (2001). 2001—A CFL odyssey : what went right ? early efforts laying the foundation. http://aceee.org/files/proceedings/2002/data/papers/SS02_Panel6_Paper02.pdf. Accessed 16 May 2016

CEE (2014). Residential lighting controls market characterization, 2014, Boston. https://library.cee1.org/content/cee-residential-lighting-controls-market-characterization. Accessed 16 May 2016

D&R International (2010) CFL market profile: data trends. https://www.energystar.gov/ia/products/downloads/CFL_Market_Profile_2010.pdf. Accessed 16 May 2016

Franceschini S, Alkemade F (2016) Non-disruptive regime changes—the case of competing energy efficient lighting trajectories. doi:10.1016/j.eist.2016.04.003

Franceschini S, Pansera M (2015) Beyond unsustainable eco-innovation: the role of narratives in the evolution of the lighting sector. doi:10.1016/j.techfore.2014.11.007

Haitz R, Tsao JY (2011) Solid-state lighting: "The case" 10 years after and future prospects. Phys. doi:10.1002/pssa.201026349

Inman GE (1939) Characteristics of fluorescent lamps. In: 32nd Annual Convention of the Illuminating Engineering Society. Ohio

Iwafune Y (2000) Technology progress dynamics of compact fluorescent lamps. Laxenburg. http://webarchive.iiasa.ac.at/Admin/PUB/Documents/IR-00-009.pdf. Accessed 16 May 2016

Navigant Consulting (2012) 2010 U.S. lighting market characterization. http://apps1.eere.energy.gov/buildings/publications/pdfs/ssl/2010-lmc-final-jan-2012.pdf. Accessed 16 May 2016

Stuart E, Larsen PH, Goldman CA, Gilligan D (2013) Current size and remaining market potential of the U.S. energy service company industry. Berkeley. https://emp.lbl.gov/sites/all/files/lbnl-6300e_01.pdf. Accessed 16 May 2016

Swope T (2010) The present and possible future CFL market. In: Northeast Energy Efficiency Partnerships

Printed in the United States
By Bookmasters